Travels in the
Genetically
Modified Zone

Travels in the Genetically Modified Zone

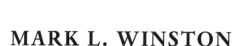

MARK L. WINSTON

Harvard University Press
Cambridge, Massachusetts
London, England
2002

Library of Congress Cataloging-in-Publication Data

Winston, Mark L.
Travels in the genetically modified zone / Mark L. Winston.
p.   cm.
Includes bibliographical references and index.
ISBN 0-674-00867-7 (hbk. : alk. paper)
1. Food—Biotechnology.   2. Genetically modified foods.
3. Agricultural biotechnology.   I. Title.

TP248.65.F66 W565 2002
664—dc21
2002017192

To Sue, for sharing her genome and the
importance of nature and nurture

And for Devora, who daily reminds us of the
profound wonder of creation and recombination

Contents

> Genetics will surely play a major role in the still infant technology of biological engineering. Already it has borne a huge harvest of practical results through improvements in breeds of food plants and animals.
>
> Theodosius Dobzhansky, *Scientific American,* 1950

Prologue

Genetically modified crops are increasingly being grown on our farms and processed into our food, and now there is a sense of urgency about resolving the diverse issues that have plagued the implementation of this new technology. The debate to date has been driven by two extreme positions, expressed by the biotechnology industry, which is seeking new products and markets, and the consumer and environmental movements, which perceive only impending medical and ecological disasters. Underlying these proponent and opponent perspectives is a compelling and revolutionary scientific innovation, our new-found ability to selectively move genes from one species to another across the boundaries that previously separated organisms.

What rivets our attention about biotechnology is this: we know that the implications are profound, but we have yet to determine whether genetically modified crops will turn out to be beneficial or harmful. We are trying to deal responsibly with the vital issues involved, recognizing that the decisions we make about tomorrow's seeds will shape our and the earth's future, but the proper way forward remains obscure.

My intention in the following chapters is not to advocate or oppose genetically modified crops, but to explore some of the issues that biotechnology in agriculture poses for society. I, like most scientists, have been enthralled by the novel and significant nature of the science, but also concerned about the proper balance between benefit and risk that we must find to safely exploit these new techniques. I also am worried about our general inability to come to grips with managing the implementation of this and other new scientific advances, and fascinated by the growing interactions between the scientific community and political, business, trade, environmental, and ethical components of the society around us. In addition, I have been astounded by the speed at which this debate has moved from the balanced perspective of compromise and conciliation toward entrenched extreme pro and con opinions.

We have tossed the genetic dice for better or worse over many millennia with crops, juggling and mixing genomes since the beginning of agriculture. Viewed historically, biotechnology comes into clearer focus as the predictable next step in our developing specificity and acuity in engineering crops. Whether this is desirable or not depends on one's perspective, self-interest, and—ultimately—one's values.

From one perspective, genetically modified (GM) crops are only an incremental change in crop breeding, inevitable after ten millennia of progress from harvesting the first crops to designing grains, fruits, and vegetables with exceedingly precise specifications. Viewed from the other side of the divide, GM crops are our greatest affront to nature yet, one that is ethically unsupportable and practically rife with risk and danger to our health and our environment.

The biotechnological advances used to produce genetically modified crops are conceptually simple. Once we understood the basic structure of the DNA that makes up genes and the chromosomes on which they are arranged, curiosity and potential benefit stimulated scientists to learn how to manipulate the genetic code.

By the 1970s we had learned how to cut a gene out of one organism, transfer it into another, and have that gene become a permanent part of the recipient's genome. The methods are not unlike those I use to transfer material from one chapter of a book I am typing to another. Just as I can cut a line from one page and insert it into another computer file, so biotechnologists can precisely remove part of the genetic code of one organism and place it into another species.

These methods are now everyday laboratory procedures, and the molecular biologists who move the fundamental components of life have become acclimatized to their techniques and complacent about the profound implications of their procedures. The public has not been as seduced by the methodological ease with which scientists can alter in days what evolution took billions of years to craft. For many, genetic engineering is a unique and dangerous turning point in human technology and civilization, taking us down paths we humans were not meant to tread.

The ethical issues posed by biotechnology are profound, but the potential benefits of gene transfers are immense. For example, the soil-dwelling bacteria *Bacillus thuringiensis* (Bt) has a gene that produces a toxin which kills butterflies and moths, many of which are important pests of crops. This insect poison has no effects on other organisms, including humans and beneficial predators and parasites of the pests. Bt has been used safely to control pests for about forty years; the bacteria are grown in fermentation chambers and then sprayed onto crops. The safety of Bt for us and the environment is rated highly enough that its use is permitted under the strict regulations governing organic farming, but its effectiveness is short lived in the field.

Scientists have taken the gene that codes for the insect-killing toxin, removed it from the bacteria, and inserted it into plants such as corn. The toxin then is produced as the plant grows, and pests that attempt to eat the plant are killed. This method of pest management has innumerable advantages, including activity of the toxin throughout the growing season, specificity for only insects that eat corn, and the concomitant reductions in chemical pesticide use and labor costs for spraying.

Other gene transfers have shown similar promise in agricultural contexts, at least to those who support genetic engineering for its practical potential. Herbicide-resistant crops are the most widely disseminated form of GM crop, containing various genes that prevent weed-killing herbicides from affecting crops. Weed control is among a farmer's most difficult and necessary management tasks. Herbicide-resistant crops allow farmers to spray herbicides while crops are in the field, reducing herbicide applications from the pre-planting and post-harvest sprays done previously to only one spray while the crop is growing.

Other types of GM crops are ready for commercial use or in development, and their potential is astounding. Frost and drought tolerance will expand the growing regions for many crops, while the insertion of nutrient-producing genes will allow foods to be used to prevent malnutrition syndromes such as vitamin A, vitamin E, and iron deficiencies. The storage lifetime for fruits and vegetables can be increased dramatically through plant biotechnology; fatty acid content for oils can be made healthier; and pharmaceutical-producing crops can be grown that will reduce the cost of producing drugs and permit people to be vaccinated simply by eating a vaccine-carrying banana, instead of having to get an injection.

Even the currently available transgenic crops have had remarkable commercial success in the countries where they have been approved for agricultural use. From 1996 to 2001, the global acreage of genetically modified crops grew from 4 to 125 million acres, mostly in the United States (68 percent of all GM crops planted world-wide) but also in Argentina (17 percent), Canada (7 percent), and China (1 percent). Soybeans, cotton, canola, and corn have been the principal genetically engineered crops with herbicide-tolerant or insect-resistant traits—or both—incorporated into their genetic makeup.

These new crops are seductive, but the allure of the methods that produce them and the benefits they provide obscures the concerns that have mobilized protest against biotechnology in agriculture and other domains. Much of this concern is focused on practical issues. Biotechnologists emphasize the pragmatic benefits of genetically modified crops, but objections expressed by environmentalists accentuate the serious risks. Innumerable press releases, Web sites, and placard-bearing demonstrators deluge us with warnings of the health and environmental dangers associated with GM crops.

We hear concerns about bioengineered genes moving from crops into the wild, borne by wind-blown or bee-carried pollen that fertilizes feral plants. We know that plants promiscuously exchange genetic material, and genes from the current generation of genetically modified crops already have been found in adjoining weeds. We're told that the proteins expressed in GM crops might harm us, inducing silent cell mutations that could erupt in cancer epidemics in twenty years or immediate and fatal allergic reactions in unsuspecting consumers. We read that GM crops might induce resistance in pest insects, forcing us back into heavy overuse of synthetic chemical pesticides.

All of these practical risks have some basis in reality, but underlying both benefit and risk is the unknown. We can speculate and attempt to predict scenarios by which biotechnology might improve agricultural production or hurt our health and environment, but obscuring these practical considerations is the realization that we are moving into uncharted territory. None but the most arrogant or foolish would claim to have a fully predictable comprehension of the consequences that will ensue from the use of genetically modified organisms, be they beneficial or harmful.

Underlying all of the practical issues is a fundamental ethical dilemma, one rooted in our individual comfort levels about how far we should go in manipulating the most fundamental building blocks of life itself. Proponents of biotechnology like to soothe these concerns by saying that what we do with gene splicing is no different from what nature has always done itself, and that we are only guiding the process with a bit more precision than we have in the past. Opponents believe just as strongly that moving a gene from a bacteria into a plant goes well beyond the natural exchanges of genetic information that organisms indulge in, and that only God for the religious and evolution for the biologically inclined have license to fabricate the design and diversity of species.

The fact is, though, that we have changed our environment phenomenally through agriculture and traditional selection and breeding even before the advent of GM crops. We have altered biodiversity dramatically, reducing crop variation to the point where bioengineering is being touted as a diversifying method that might save us from descent into crop homogeneity. We have almost eliminated the forerunners of today's crops in the wild, and must substitute genes from bacteria or animals for the natural genomes we have destroyed. We produce food that contains pesticide residues, a health risk that may be considerably more dangerous than the unknown effects of genetic modification.

GM crops may be regarded as just another step in the improvement of agriculture, with environmental and human health consequences of a magnitude similar to those we have come to accept. Or, biotechnology might differ dramatically from past methods, posing unprecedented ethical issues and health and environmental dangers that are considerably more problematic than any current practice.

In some ways the debate about GM crops involves ethical and practical themes that have plagued us throughout the development of agriculture, especially issues regarding the balance between benefit and risk that we confront with all new technologies. In other ways the idea of GM crops is unique, because to create them we rearrange genomes in months, whereas natural selection took almost four billion years to induce the current arrangement of genetic material in the earth's biological realm.

My own interest in and attitudes about biotechnology have been influenced by my dual pastimes as scientist and writer. I continue to oversee an active research program investigating bee biology and management at Simon Fraser University in Vancouver, British Columbia. Although my laboratory's studies do not involve producing transgenic organisms, the role of bees as crop pollinators provided a personal link to the revolutionary changes in contemporary agriculture that are being driven by biotechnology.

Bees forage on the nectar and pollen produced by many crops, including plants, such as canola, cotton, and soybeans, that are increasingly being genetically modified. Beekeepers began asking me in the mid-1990s whether these crops might harm bees, and I contacted various U.S. and Canadian government regulatory offices as well as their counterparts in private industry seeking data I could examine to address the beekeepers' concerns. To my surprise and considerable annoyance, I repeatedly was refused access to the studies themselves for "proprietary" reasons, although I received undocumented assurances that GM crops were safe for bees from both government and industry representatives.

Research since conducted by independent university scientists, as well as some studies done in my own laboratory, has reassured me that at least the current generation of GM crops are safe for bees. However, this irritating entrée to the politicized world of agricultural biotechnology piqued my curiosity about the broader issues associated with genetically modified crops. I soon began to see genetic engineering as emblematic of a bigger problem, our inability as a society to come to grips with the already impressive but still growing impact that science has on our daily lives and decisions.

I also am a commentator attentive to and absorbed by the interaction between scientific advances and societal issues, and found myself fascinated by the irrational, rhetoric-filled debate surrounding genetically modified crops. As societal discussions about bioengineering heated up, I became as confused as any member of the general public about the potential impact of transgenic crops on human health, environmental integrity, and the way we understand our place in the global context of the other organisms with which we share our fragile planet.

I spent more than two years traveling around the United States, Canada, and Europe talking with proponents and opponents to learn more about the issues GM crops pose for the public, and about how the science behind biotechnology is perceived from diverse perspectives. I visited high-security laboratories and hole-in-the-wall environmentalist storefronts, walked through farms and sat in innumerable offices, and searched through cyberspace for documents, information, and contacts. I tried to keep from forming any unduly hardened opinions, and by keeping my own perspective fluid I found myself sympathetic to almost every point of view I encountered.

In the following chapters I have tried to capture some of the mood and feeling that pervades each camp. My objective is not to prove that GM crops are good or bad, for in fact they can be both. Rather, I want to convey the sense of excitement, curiosity, and triumph that permeates industrial, government, and university laboratories inventing genetically modified crops, and the panic and gut-wrenching opposition that fills the minds and offices of legions of dedicated environmentalists and consumers.

I hope to evoke some of the desperation that characterizes conventional farmers today, their trust in scientific solutions for broader agricultural issues, and their hope that biotechnology will solve many of the economic and production dilemmas that are driving many farmers off their farms. In contrast, I present the anti-GM philosophy and practices of productive and successful organic farmers, and the intense fear they harbor that GM crops will spread into their fields and destroy their rapidly expanding industry.

Traveling on paths parallel to conventional and organic farmers in developed countries are third world farmers, caught in powerful trade and economic vises that allow for little flexibility in making their own decisions about biotechnology. Also moving through these pages will be regulators attempting to come to grips with setting boundaries for the commercialization of biotechnology, lawyers wrestling with issues about patenting the essence of life itself, and advocacy groups of every persuasion using increasingly sophisticated tools to manipulate public opinion.

I will, at the end, weigh in with my own conclusions, but I will share one insight into my underlying bias right at the start. We North Americans live in an economically rich and prosperous society, but our ability to listen to each other, and to understand the diverse perspectives that we bring to the most important issues facing us today, is impoverished. The best choices usually lie somewhere in the middle ground between technological benefit and environmental risk, but debates today have ceded the center and gravitated to the extreme edges. The rhetoric and posturing will diminish only when we accept the validity of each other's points of view, a position that will naturally lead us back to the middle ground.

Listen to all the voices, lay aside your preconceptions, and reflect on the marvelous diversity of opinion and feeling that are our greatest resource.

> The skills of farmers are centered not on their relationship to the world but on their ability to change it.
>
> **Hugh Brody,** *The Other Side of Eden,* **2000**

Seeds

On 23 September 1959, during a rare trip to the United States, Nikita Khrushchev made political and agricultural history by visiting the Coon Rapids, Iowa, farm of Roswell Garst. Khrushchev was then premier of the Union of Soviet Socialist Republics, and his trip took place during the most frigid portion of the politically icy Cold War between the communist and capitalist worlds.

Khrushchev wanted to do only two things during his trip besides meet with U.S. President Dwight D. Eisenhower: visit Disneyland and shop in Coon Rapids. He did fly out to California, but the U.S. State Department caused a diplomatic incident by forbidding him to go to Disneyland for "security reasons." Fortunately for the Soviet Union his shopping trip to Coon Rapids was more successful. Khrushchev was in the market for seeds, and Roswell Garst was ready to sell not only his hybrid corn seed but a holistic philosophy that had revolutionized American farming and quadrupled U.S. corn production.

Garst and Khrushchev had become acquainted during the mid-1950s after Khrushchev mentioned in a speech that the United States had a good agricultural system and that farming could be the basis for developing a better relationship between the antagonistic superpowers. After that Garst visited the Soviet Union a number of times, and then Khrushchev came to Iowa to experience the Garst farm for himself.

The USSR was starving because the Stalinist regime that preceded Khruschev's had invested heavily in industrial development and military power but had ignored agriculture. Khrushchev and Garst were united by their mutual passion for farming and their shared vision that the cultivation methods and perspective developed in the United States could make Soviet agriculture as productive as American farms. They hoped that cooperation in agriculture not only would help feed the hungry communist world, but also would bridge the political divide that kept both superpowers on the brink of conflict and nuclear destruction.

Khrushchev's visit to the Garst farm was a particularly compelling moment, but it was just that, a brief episode that was preceded and followed by decades of astounding changes in farming. In the years leading up to Khrushchev's trip, the influence of Roswell Garst on American agriculture had been profound. Garst had begun by selling technologically innovative hybrid seed corn, but had gone much further by using that seed to create and promote an agricultural system that produced unprecedented yields. Farming in the United States was changing, improving not only through technical advances but through a uniquely American fusion of farming and industry, a movement in which Garst was the leading figure.

The creation of hybrid corn led to the development of an agricultural empire based on strong corporate involvement, an empire that eventually matured further into the multinational corporations that are today inventing and selling genetically modified crops. In Garst's era, scientific advances were adopted by agribusiness and had considerable influence on global politics. Today, scientific discoveries are having a similar effect, with biotechnology affecting not only crops themselves but the way we conduct the business of agriculture, international politics, and trade, and our ability to feed a growing global population.

In farming, then, now, and always, it begins with the seed.

■■■ A seed is a marvelous entity, expressing a tiny moment of evolutionary history. Its ancestors evolved through billions of years, beginning with the most primitive bacteria, evolving on past the earliest plantlike organisms to the seeds of contemporary and complex plants. Soon it will fall to the ground and sprout, producing its own offspring, which will continue the evolutionary dance through endless years of future birth and death, its descendants' traits mediated by the ongoing forge of natural selection.

Evolution and natural selection are without purpose or goal, but they have inexorably increased biodiversity. First kingdoms, then phyla, and on through species, the numbers and types of organisms have grown steadily over time. Occasionally there has been a catastrophe of epic scope that diminished diversity to a shadow of its former exuberance—a meteor hitting the earth, a volcano darkening the globe, tectonic shifts submerging continents. Inevitably the biological world rebounded, with new groups of species slowly diversifying until the next natural disaster.

Now we have imposed a new twist—biotechnology—on this pattern of fluctuating diversity. Today we are selectively, deliberately, and surgically reinventing the plant world, rearranging genomes in a matter of months or years, at rates astronomically faster than natural selection itself or even traditional plant breeding could produce new varieties. Contemporary seeds are increasingly designed by humans, with implications of the profoundest magnitude for both agriculture and what remains of the wild world.

Our current ability to precisely engineer crop genomes was

preceded by a long history of genetic manipulation in agriculture. Human impact and its accompanying effects began early in our history, at many tropical and subtropical sites around the globe. Our ancestors were omnivores, fortuitously consuming whatever plant or animal material they encountered. Even then humans had considerable effects on the environment, reducing and even driving to extinction populations of the animal species they hunted and expanding the distribution of plants by accidentally distributing seeds as they migrated.

Humans probably first realized that seeds could yield a stable food supply through agriculture when they observed plants arising from refuse or wasteland, perhaps fruit trees growing along forest and jungle paths from discarded or defecated seeds, or vegetables sprouting in garbage dumps at temporary settlements. A more organized approach to agriculture began about eight to ten thousand years ago, coincidentally at a number of locations around the globe. The most diverse farming developed in the Near East, with legumes, cereals, flax, sesame, and fruit trees. At about the same time, New World residents were growing beans, maize, squashes, and potatoes, and Asian farmers were beginning to cultivate rice.

These early domesticated crops foreshadowed the overwhelming changes contemporary agriculture has wrought in plants. Humans soon learned to separate crop varieties from wild types, in order to prevent characteristics undesirable for cultivation from mingling with those selected for farming. Continued selection of crops with desirable characteristics increased the separation between feral and managed plants, and accelerated the diminishing diversity and more limited variation found in today's crops.

The simplest way to select crops is to save seeds preferentially from plants with beneficial traits, and the first farmers selected for large seeds and fruit, increased seed production, lack of dormancy, faster germination, higher annual yield, and reduced seed scattering. The success of this early selection resulted in an accelerating impact of agriculture on crop diversity and feral plants. Crops quickly became commodities, moved and traded over a rapidly widening area, so that many plants were distributed well beyond their previous ranges, and some throughout the globe.

Three phenomena have characterized the more recent impact of agriculture on the earth. The first was the increase in human population, which has doubled at shorter and shorter intervals over the last thousand years. The result was increased acreage under cultivation and a fundamental remodeling of the globe toward managed rather than wild ecosystems. By 1998, 3,410,523,800 acres of land were under cultivation worldwide, an area larger than the United States. Entire ecosystems have disappeared, others remain but are threatened, and the sheer volume of people and area of farmland has been a major force of biological change.

The second event through which agriculture modified our planet was European colonization. Previously, migration and trade had moved crops between countries and continents, but the Europeans inaugurated an unprecedented dispersal of biological material worldwide. Corn, tomatoes, and potatoes were transported from the New World to the Old; wheat, rye, and barley carried from the Old World to the New; and rice, soybeans, and alfalfa moved from their Asian sources to every arable continent. Each of these and innumerable other introductions conveyed not only unique genetic material, but also assemblages of introduced plant pests and diseases that today cause the majority of pest-management problems around the world.

The third factor shaping the nature of agriculture and the environment alike is the increasing precision with which we have selected and bred crops. This acuity stemmed from many advances, but at its heart lies the work of two men, one the English naturalist Charles Darwin and the other an Austrian monk, Gregor Mendel. The concepts of evolution and genetics were not their work alone, but both of them were decades ahead of their colleagues in synthesizing the companion concepts of natural selection and inheritance that are at the core of all contemporary biological science, and that form the substrate upon which biotechnology grew.

In retrospect, it is remarkable how dramatically our farmer ancestors changed crop plants through selecting desirable varieties, given how little they knew about natural selection and heredity. We now know much more about the concepts underlying variation, and particularly about the extent to which many plants can recombine their genetic material and absorb genes from related species. Critics of genetically modified crops may find it uncomfortable to contemplate, but plants are naturally fluid in their propensity to blend genomes. Our ability to mix and match genetically controlled traits artificially had become extensive even before Darwin and Mendel, but the twin pillars of evolution and genetics accelerated advances in horticulture and led to unprecedented advances in varietal selection and breeding.

Darwin's impact was felt first. Although we remember him primarily for his role in expressing the concepts of evolution and natural selection, his work also had a transforming impact on crop breeding. Indeed, he began *The Origin of Species* with a chapter titled "Variation under Domestication," reflecting the considerable influence that crop, pet, and livestock breeding had on his thinking. Darwin noted that plants show extensive variation in nature, and suggested that much of this expressed variability is pruned out by natural selection. Farmers, however, can propagate and nurture desirable traits that might not survive in the wild but can thrive under cultivation.

Domestication demonstrated for Darwin the inherent simplicity of choosing successive generations of gradually improving stock that over time could transform inferior plants into functional crops: "No one would expect to raise a first-rate melting pear from the seed of the wild pear . . . The pear, though cultivated in classical times, appears from Pliny's description to have been a fruit of very inferior quality. I have seen great surprise expressed in horticultural works at the wonderful skill of gardeners, in having produced such splendid results from such poor materials . . . The gardeners of the classical period, who cultivated the best pear they could procure, never thought what splendid fruit we should eat, though we owe our excellent fruit . . . to their having naturally chosen and preserved the best varieties they could anywhere find."

Crop selection had been operating under Darwinian principles for millennia, but it was Darwin's articulation of the inherent and extensive variation within species, the nature of heredity, and the similarities between artificial and natural selection that provided a clear explanation for what farmers had been doing intuitively since the beginning of agriculture. Perhaps the most remarkable aspect of Darwin's theories was that he was not aware of the physical mechanisms of genetics responsible for determining heredity. As he put it, "The laws governing inheritance are quite unknown," and although the key breakthrough was published in 1866, the laws of heredity were to continue languishing in obscurity until the early 1900s.

Gregor Mendel added the second insight leading to scientific plant breeding and eventually biotechnology with his research demonstrating the particulate nature of inheritance. He, too, was missing an important part of the puzzle, since the composition of the "particles" was not fully explained until the second half of the twentieth century. Nevertheless, his careful breeding experiments with peas provided the framework upon which the science of modern genetics grew.

Mendel was the son of an Austrian farmer, ordained as a priest in 1847 and later made abbot at the Augustinian monastery in Brünn, Austria-Hungary. Although he taught science, Mendel was a poor student, and never passed his formal examinations in spite of many years of his own remedial schooling. He worked alone, spending close to ten years crossing peas and observing generations of offspring.

Mendel examined many traits, including the easily visible characteristics of seed shape, plant height, and floral color, and noted that traits seemed to skip a generation in being expressed when plants of different types were crossed. He calculated the mathematics of the inheritance patterns he observed, and determined that each trait was dominant or recessive, inherited independently, half from the father and half from the mother plant. These concepts inevitably led to the conclusion that heredity involved physical particles, which we now know as genes and the chromosomes that carry them.

Mendel's work was published in scientific journals, and presented at a number of meetings, but oddly was lost to mainstream science until 1900, when three other European botanists independently arrived at the same results when experimenting with other plant systems. All three rediscovered and cited Mendel's work when searching the literature before publishing their own research, and thus he justly received credit for discovering the physical nature of heredity.

Scientists soon understood that heredity was determined by particles within cells, and that chromosomes composed of DNA carried these particles, or genes. At this point all the components were in place for contemporary approaches to plant breeding, including an understanding of evolution, selection, and the nature of inheritance, as well as the ability to predict results from designated crosses between varieties. With these conceptual and practical tools, plant breeders made enormous strides in developing new crop varieties.

In no crop was this revolution more noticeable than in corn, and no one was a better promoter of linking science and business than Roswell Garst.

■■ Iowa today has not changed much from the way it was when Garst was born and raised. It is still a place where farm values predominate and agriculture, especially corn, drives the economy. It may be the most managed environment on earth, with almost all of its land under cultivation. Slightly rolling terrain is dotted with clusters of farm buildings, typically including a modest wooden house shaded by a few old trees with more elaborate barns and silos grouped nearby, all surrounded by cultivated fields extending as far as the nearest neighbor's farm. The land is dotted with farm ponds and the occasional lake, created over the centuries by damming creeks and rivers for irrigation and flood control. Plumes of dust rise up from spring to fall, formed by the ubiquitous tractors and combines that move through the fields as farmers go about their business.

Most adults in Iowa farm or work in the multiple industries that support farming. Overalls and jeans are more common than suits and ties, and many people wear caps embossed with the names of seed companies, agricultural chemical firms, and manufacturers of farm equipment. Iowans tend to be gregarious, and often sprinkle their conversations with old-fashioned phrases like "sure as shootin'." Towns greet incoming visitors with signs proclaiming slogans such as "Watch Us Grow" and "A Community United for Progress," and this mood of optimism is deeply and permanently ingrained in the state's psyche.

Roswell Garst was born into this milieu in 1898, in the central Iowa town of Coon Rapids on the middle fork of the Raccoon River. Today this is the only free-flowing river remaining in the state, and the town still is proud of the unpretentious rippled rapids for which Coon Rapids was named. The edge of the Wisconsin Drift Glacier had its southernmost limit there and the surrounding area is hillier than most of Iowa, with creek valleys cutting through the hills and more wildlife than in the flatter and cultivated terrain that makes up the rest of Iowa. The nearby farmland is especially fertile, still full of the minerals and organic matter deposited when the glacier slowed and came to a halt.

The land and people seem stable in a permanently managed landscape with a perpetual set of values, but some things have changed since Roswell Garst's era. Since 1929, corn yields have increased from 26 to over 130 bushels per acre, and he and Garst Seeds, the company he founded, deserve much of the credit for that remarkable statistic. Along with that increase in production have come radical changes in the business of farming, with increasing corporate involvement leading to farmers' dependence on industry to provide their seeds, fertilizers, and pesticides.

Because of these changes, corn became one of the world's three leading food crops, along with wheat and rice, but its ancestral background is obscure. The difficulty in tracing corn's origins is that there are three closely related species complexes—those of maize (*Zea mays*), teosinte (*Euchlaena mexicana*), and tripsacum (numerous *Tripsacum* species)—still growing wild in Mesoamerica. One theory suggests that all three had a common and now extinct ancestor, while others have proposed that maize descended from teosinte, or teosinte from maize. Most likely, maize and teosinte crossed and back-crossed with each other hundreds or thousands of times, maintaining their identity as species but naturally exchanging genes and evolving by incorporating each other's advantageous characteristics.

Maize cultivation began about seven thousand years ago in what is now Mexico and the southwestern United States. Scientists refer to the cultivated types as "corn" to differentiate them from their wild ancestors, and the characteristics of this early corn were remarkably different from those of the types that were eventually selected and bred. Small cobs no larger than a thumb were typical of the first cultivated crops, with kernels containing relatively tiny amounts of starch. Over time, local varieties of cultivated corn with larger and more nutrient-packed kernels began to appear in farmers' fields, gradually differentiating into flint, dent, and sweet corns, the ancestors of contemporary varieties.

The diversity of corns cultivated by Native American cultures was astonishing, based as much on mystique and religious overtones as it was on agronomic characteristics. The native approach to growing corn focused as much on maintaining varietal purity as it did on yield and ear qualities. The color and type of kernel had religious meaning as well as food value, and agricultural practices that preserved varietal integrity were of prime importance. Thus fields of each variety were planted separately to prevent the cross-pollination that would infringe on varietal purity.

As a result, hundreds of corn varieties were being grown when Europeans first arrived in the Americas, and while not generally cultivated today they can still be found in seed banks maintained by the U.S. Department of Agriculture at Iowa State University in Ames. These seed banks contain over seventeen thousand varieties of corn stored as dried seeds in clear plastic containers under cold conditions, some collected from wild corn ancestors, others from the cultivated aboriginal varieties, and the remainder representing thousands of selected and bred varieties from the last three hundred years of corn farming.

Just the physical differences visible in these varieties is extraordinary. Kernels of the smallest varieties are as small as poppy seeds, while the largest kernels are bigger than lima beans. Cobs vary in size from the length of a finger to as long as an arm, from pencil-thin to as thick as a farmer's forearm; some are perfectly round and as small as a golf ball and others are shaped like a ripe strawberry. Kernel color in the banked varieties is equally diverse, ranging from brown to yellow, purple to green, orange to black, with some cobs spotted with many different colors.

European settlement and the accompanying growth of mass agriculture changed the aboriginal approach that naturally maintained this variation. Farmers moved away from genetic diversity and toward fewer varieties by selecting only for plants that produced the most kernels of the largest size with the greatest sugar and starch content and a predominantly white or yellow color. At first, this corn breeding mixed up genomes from the previously intact varieties, creating more genetic variation within a corn seed but losing the boundaries between varieties that had maintained that variation in an organized manner.

Scientists began to inbreed corn around the beginning of the twentieth century, searching for improved and more homogeneous varieties, eventually inbreeding so severely that a field of inbred corn could not be fertilized by pollen from the same variety. The new highly inbred lines were poor producers and breeders on their own, but they could be crossed to other inbred stock to produce hybrid corn. The advantage of this convoluted selection scheme was that hybrid corn, like most hybrid crops, produced unprecedented yields, in some cases three to four times those of the ancestral stocks.

By 1930, improvements in scientific breeding methods had created some exciting new lines of hybrid corn. The stage was set for an entrepreneur in the person of Roswell Garst to meld salesmanship with science, beginning the transformation that led to today's multinational biotechnology companies.

Genetically modified crops would not have been successful in 1930 even had the methodology been available, because agribusiness infrastructure and farming practices were not yet ready for biotechnology. Roswell Garst made immense contributions to the agriculture of his time, and his life's work was a necessary precursor to transgenic crops in our time. He and his compatriots changed the lifestyle of agriculture: they transformed individual self-reliant farmers into a community of growers highly dependent on multifaceted companies that supported high-input agriculture, thereby providing a corporate template through which genetically modified crops could enter the marketplace.

Garst persuaded farmers to stop saving seeds every year and instead to purchase their seeds from companies that had selected more productive varieties, creating a dependency on seeds produced by corporations, new crop varieties to which genes eventually would be transferred, and a sales network through which genetically modified seeds would be sold to farmers in our generation. American farmers also began producing excess food as a result of Garst's influence, and this surplus combined with increased input costs created an economic environment that is driving agriculture toward biotechnological solutions to the problem of a collapsing farm economy. Garst also made international trade arrangements with developing countries that today remain as possible conduits for genetically modified crops. Finally, he was confident that agricultural problems could be solved, an attitude that in many ways was his most important legacy to what eventually became agricultural biotechnology.

Garst grew up in a farming community, and given his background, enterprising nature, and intuitive feeling for agriculture, it is not surprising that he had become a successful dairy farmer by the 1920s. He was restless, however, and when his father died and subsequently the family's general store was sold, Garst moved to Des Moines, Iowa, to subdivide and sell 120 acres of former farmland at the city limits.

Garst and his new wife, Elizabeth, were well connected to Des Moines society through their family's long history as successful merchants and farmers. They soon began moving in the same circles as Henry A. Wallace, who was from a highly distinguished third-generation Iowa family and would later become U.S. Secretary of Agriculture under Franklin D. Roosevelt and eventually make an unsuccessful run for the Presidency. Wallace was then editor of *Wallace's Farmer*, the most influential farm newspaper in the Midwest, and on the side he had developed a serious interest in corn breeding.

Wallace perfected techniques to produce the higher-yielding hybrid corns by detasseling male plants and thereby preventing undesired crosses. Corn has its male pollen-producing structures located in the tassels at the top of the plant, and the female part that produces the ears of corn is located farther down on the stalk. Wallace would cut off the tassels of one variety planted next to a row of a second variety, on which he would leave the pollen-producing tassels intact. This procedure insured that pollination would occur only between the tasseled variety and the other detasseled line, resulting in outcrossing between inbred varieties and the higher-yielding hybridized corn.

Garst quickly grasped the sales potential of these superior-yielding varieties, and offered to sell Wallace's hybrid corn to farmers as seed corn from which to grow their crops. The arrangement they made was that Wallace's new company, Pioneer Hi-Bred, would select and breed the inbred lines, and provide those to Garst under license; then Garst's new company, Garst Seeds, would plant this foundation stock in sufficient quantities to produce enough hybrid seed corn for farmers to purchase the following season.

By 1930 the Great Depression had hit, Garst's Des Moines subdivision was moribund, and he and Elizabeth had moved back to the family farm in Coon Rapids. At the time corn farmers were not accustomed to buying seed, or in fact to buying almost anything. Typically Iowa farmers purchased only salt and nails, and were otherwise self-contained, saving part of their crop for next year's seed, using manure for fertilizer, and relying on draft horses to plow in the spring and pull trailers in the late summer, into which they threw the hand-picked and husked corn they harvested.

Selling seed corn to this clientele would have been tough at any time, but during the Depression it seemed an impossible task. Garst hit on an ingenious strategy, giving farmers free bags of corn in the spring and asking only that they pay him half of the increased yield from his hybrid corn in the fall. When his superior varieties out-performed the farmers' seeds, Garst accepted only the cost of the seed corn he had given the farmers in the spring, along with their commitment to buy seeds from him the following season. He also created a sales network, with the most respected farmers in each county across the Midwest as his representatives. Within a decade most farmers had switched to buying seed with cash.

The same sort of thing happened with many contemporary crops: a few varieties are selected by experts, maintained by corporations, and sold to farmers. Hybrid corn is clearly more productive, but farmers cannot maintain these lines themselves. The parent inbred lines are too complex to produce and maintain without considerable scientific input and an elaborate breeding program. Further, hybrid corn is genetically handicapped by a diminished ability to grow viable, productive plants in future generations, so that saving hybrid seed is not sensible.

As a result, corn farmers ceased selecting and saving their own seeds and came to depend on newly developing corporations such as Garst Seeds that specialized in hybrid seed corn production. This system was integral in improving productivity, but it also changed farming from a simple occupation to an industry, and severely reduced the number of varieties being produced for commercial agriculture. In the United States, for example, 786 corn varieties were available in 1903, but only 52 in 1983, a decrease of 93 percent. Thus farmers who became used to purchasing the latest hybrid variety from their favored seed company each spring were ideally predisposed to accept transgenic corn when those varieties became the corn of choice recommended by seed companies in the mid-1990s.

Not only did farmers become dependent on agricultural conglomerates for seed to plant each season, but they came to rely on those companies for an increasingly essential array of pesticides to combat pests, weeds, and diseases, synthetic fertilizer to reach the yield potential contained within hybrid seeds, and increasingly specialized and sophisticated farm machinery to harvest the more uniform corn plants that lent themselves to mechanical picking and processing. Garst invested in all these areas, inventing new machinery and promoting the benefits of heavy synthetic fertilizer applications that interacted with hybrid corn to further boost production.

Agribusiness had been born, integrating seeds, fertilizers, pesticides, and farm machinery into a corporate structure, with suppliers linked by increasingly complex business relationships. Hybrid seeds quickly evolved beyond a product and into a system, a way of farming that produced exceptional yields and soon came to the attention of Nikita Khrushchev.

The Khrushchev visit was important for the eventual development of the agricultural biotechnology business in both its direct outcomes and in the uniquely American attitudes it reflected that led to genetic engineering. Khruschev's trip set U.S. foreign policy moving squarely toward favoring agricultural aid as an important instrument of diplomacy. It also alerted American farmers to previously underutilized marketing opportunities overseas for U.S. crops. In addition, the North American perspectives on farming, food, and corporations that were then becoming prominent continued to develop, putting biotechnology on a collision course with agricultural, economic, and ethical values in other regions of the world.

Khrushchev and Garst were similar in many ways, both coarse, earthy, colorful, and rotund men who enjoyed a joke and were unabashedly passionate about farming. Khrushchev was admittedly envious of American success in producing food. He took serious political risks at home by acknowledging U.S. supremacy in that area and encouraging Soviet bureaucrats and farmers to learn how to farm from the enemy. Roswell Garst was a confident capitalist committed to making money, but also driven by a deep belief that the political divide between communist and capitalist societies could be bridged through farming. His core philosophy was that by creating international markets Americans could help others, reduce political tensions, and stimulate trade and profit for all.

Garst took a considerable amount of political heat from his American customers for reaching out to the communists, and some of them even canceled their seed orders. I spoke with his granddaughter Elizabeth (Liz) about Khrushchev's visit, which impressed her because of the hordes of spies, helicopters, National Guardsmen, reporters, and food tasters that flocked to Coon Rapids, turning what Khrushchev and Garst had hoped would be a visit between two farmers into an international event.

Like most of Roswell's children and grandchildren, Liz Garst left Coon Rapids to get an education, in her case to study agricultural economics and business at Stanford, Michigan State, and Harvard universities. She then joined the Peace Corps and spent some years in Colombia, followed by a few years at the World Bank offices in Washington, D.C.; in 1991 she returned to Coon Rapids to help run the family farm, machinery, fertilizer, and bank businesses that sprang from the original parent company, Garst Seeds. We talked at one of the family homesteads, on a porch overlooking the Raccoon River watershed, which the Garst family privately owns and is maintaining for posterity as undeveloped and unfarmed parkland.

She told me about her grandfather's purpose in cozying up to the communists: "Roswell was frequently accused of being a commie sympathizer but he was so into capitalism that he would even sell to dirty dog communists. He was trying to make money and proud of it. His other motive was that he agreed with his mentor, Henry Wallace, that the arms race was crazy and that hungry people were dangerous people." Khrushchev concurred; he startled the more than fifteen hundred members of the press who had descended on the Garst Farm on the day of his visit by saying, "Let there be more corn and more meat and let there be no hydrogen bombs at all."

LIFE magazine in its cover story about the visit frivolously called it "A Cornball Act Down on the Farm," but the outcome was important. Garst immediately petitioned the U.S. State Department for an export license to sell hybrid seed corn to the Soviet Union, an astounding request because at the time there was no trade between the two countries in any commodity whatsoever.

He was persuasive, and the new Democratic government of John F. Kennedy that came into office in early 1961 was intrigued by Garst's advocacy of agricultural trade as a way to influence the Soviet system. The seed sale was approved, and was followed by sales of farm machinery and fertilizer that contributed to the doubling of Soviet corn production in only two years. Further, the Kennedy administration made farm aid and international trade a cornerstone of its agricultural policy, opening up markets for U.S. farmers, building international relationships between corporations and governments, and promoting the American vision of high-production agriculture for which today's genetically modified crops are ideally suited.

Khrushchev came to Iowa primarily to learn about productivity increases from the Garst system, but also because he was impressed with Garst's get-it-done forthrightness, considered by many a characteristic American trait. Garst was a spell-binding speaker and indomitable advocate of the American way of life. In one filmed interview that Liz Garst showed me, Roswell spoke forcefully about spreading American agricultural know-how internationally to contribute to feeding the growing world population: "It's the American way of doing it, and always will be, we hope. I think we have work to do, and it's going to be fun doing it."

This attitude holds out the promise of achieving prosperity by using scientific advances to increase agricultural production and by accepting heavy corporate involvement in farming. Biotechnology is the logical extension of this philosophy, and many Americans have been surprised at the resistance encountered elsewhere in the world to further expansion of the multinational U.S. agribusiness empire through the utilization of genetically modified crops.

The biotechnology industry has suffered from its lack of a spokesperson as inspirational, articulate, and trustworthy as Roswell Garst. I spoke with his son David about his father's personal presence: "I'd say that my father was a man of unbelievable vision and also unbelievable simplicity and salesmanship. My father was very astute at taking complicated ideas and making them so simple that everybody could understand them. He could paint the cloud and then he could make the sun shine. He filled people's minds with a vision of prosperity, a vision of progress, and visions of humanity."

Nikita Khrushchev and Roswell Garst had a vision of prosperity through agricultural innovation, but what happened in the next generations evolved in directions they never could have foreseen. Khrushchev eventually was driven from power, and the Soviet economy was not robust enough to buy the expensive fertilizer, pesticide, and machinery needed to take full advantage of hybrid corn and other crops. Farming in the Soviet Union collapsed in the late 1980s, and from then until today the countries in the region have had to import massive quantities of grain to stave off starvation.

Garst Seeds and the other seed-producing companies also experienced dramatic changes. Roswell Garst died in 1977 and the Garst family lost control of Garst Seeds soon thereafter, ironically because of industry trends caused by the successful farming philosophy he espoused. By the early 1980s both the farmers and the corporations that supplied them were discovering that productive high-input agriculture was not economically sustainable even in the United States. Harvests were way up for all crops, and high yields had forced corn prices down from $3.11 in 1980 to $1.90 a bushel in 1999. At the same time, fertilizer, pesticide, and fuel costs skyrocketed, so that today farmers pay out $1.00 in expenses for every $0.40 in income they receive from selling their crops.

This lethal combination induced intensive intervention by the federal government to shore up this increasingly uneconomic system. Today up to 50 percent of American farm income comes from subsidies paid directly to farmers by the government, including payments for not planting their fields. It is only because of this federal support that American farming avoided the collapse experienced by the Soviets, but U.S. agricultural policy today is struggling to balance the economic viability of farming with international pressure to reduce tariffs and subsidies on farm products.

Agricultural chemical companies also began experiencing diminishing profits, caused by a combination of higher production costs for pesticides and fertilizers, the increased investment necessary to persuade government regulators that their products were safe, and the diminished ability of farmers to purchase their products. These corporations had become multinational, but they chose a typically Garst-like response to this economic crunch. They decided to invent their way out of this economic dilemma by embracing biotechnology.

Virtually all of the agricultural chemical companies began to recreate themselves as life sciences companies in the early 1980s, on the basis of their perception that GM crops would be more profitable than the hybrid crops and could be grown without the problems that developed with hybrid seeds. For one thing transgenic crops were viewed as having an excellent potential to reduce the need for the increasingly unprofitable pesticides and fertilizers that had originally been the basis of their businesses. For another, corporate decision makers rightly or wrongly thought transgenic crops would be environmentally cleaner than hybrid crops grown with chemical pesticides, and they believed that biotechnology would reduce the need for expensive regulatory compliance and improve the acceptability of their products to the public. Finally, corporations were seduced by the potential of this new technology to create entirely new markets for crops with novel nutrient profiles or plants that would be pharmaceutically useful.

All of the agrochemical companies initiated massive research efforts to produce GM crops during the 1980s, but they also quickly realized that they were missing one key component for a successful biotechnology strategy. They needed seeds in which to put the genes they wanted to splice into crops. Thus they began to collaborate with and then buy up seed companies in order to guarantee their access to seeds.

What happened to Garst Seeds was a direct result of these decisions throughout the agribusiness complex. Today Garst Seeds is fighting for its corporate independence, having been bought, merged, remerged, and eventually spun off as a result of biotechnology. The process began when the company's former supplier, Pioneer Hi-Bred, offered to purchase Garst Seeds in the early 1980s. Roswell Garst's descendants did not want to sell, so Pioneer canceled the rights of Garst Seeds to use Pioneer inbred lines to produce commercial seed. The Garst company was left without seeds to sell and frantically attempted to establish its own research and breeding program, including the development of genetically modified crops. However, in 1986 the need for considerable and immediate financing forced the Garst family to sell their company to Imperial Chemical Industries (ICI), a chemical firm hoping to become a leader in agricultural biotechnology.

ICI also overreached, however; it could not fund the massive and decades-long research effort needed to create transgenic crops, and in turn had to split off its life sciences divisions into Zeneca, then Astra-Zeneca, and finally into a merger with Novartis to form Syngenta (see Chapter 2). Garst Seeds was purchased and merged along with its parent companies, but by 2001 Syngenta owned nine other seed companies and had spun Garst Seeds off into an independent entity.

Through these intricate machinations, Garst Seeds became the only substantial seed company existing outside of a conglomerated corporate umbrella. Pioneer Hi-Bred was bought by DuPont in 1999, and Monsanto purchased the next largest independent seed company, DeKalb Seeds, in 1998. Similarly, Dow Chemical now owns Cargill Hybrid Seeds, United Agriseeds, and Illinois Foundation Seeds, and today's two other major life sciences companies, Syngenta and Aventis, own the remaining major seed companies between them.

The Garst company's independence may not last, however. Rumors abound of an impending sale to one of the current leading biotechnology multinationals or to the companies BASF or

Bayer, both of which are major international chemical companies that need a seed outlet if they are to join the other biotechnology companies in the quest for transgenic profits.

█████ Somehow Garst Seeds retained Roswell Garst's influence throughout all this buying and selling. The corporate headquarters moved east to Slater, Iowa, but it still feels like a place where a farmer can walk in, have a cup of coffee, put his boots up on a desk, and exchange the latest farm gossip. Optimistic company slogans dominate the modest entrance to its functional, one-story research facility, testifying to the company's attempt to carry on Roswell's marriage of farming and corporate cultures: "Bringing technology to your field," "We do business face to face," "We combine modern science with traditional values," and "Today is the future we built long ago, on the foundations of our beliefs, values, and dreams."

Everything still comes back to the seed. Roswell Garst understood that you could start with an inventive approach to seed selection and breeding, and build on that to create an agricultural empire with interdependent corporate nodes touching on every component of farming. So it was for hybrid corn seeds and so it has begun for biotechnology, with the creation of transgenic seeds now accelerating the move toward an even more complex, multinational, and corporate way of doing farming.

Independent seed companies, side by side with fertilizer, pesticide, and farm machinery industries, became the way of doing agriculture in Garst's era. Today, this corporate structure has been bought up and superseded by the bigger, better-funded, and voracious biotechnology industry that is looking for seeds in which to place their genes and a sales network with which to sell them. Companies believe that these changes will reduce the use of pesticides and fertilizers, the former mainstays of corporate profits, but the biotechnology industry is betting that profits from transgenic crops will replace that income.

Of course, not everyone agrees with this vision, and the opposition to genetically modified crops is rooted as much in alternative conceptions of farming and an aversion to corporate control of agriculture as it is in the concept of the crops themselves. As we shall see, many critics of transgenic crops argue that we should invest in devising methods of sustainable low-input farming rather than in biotechnological advances they view as perpetuating the current input-dependent system of agriculture. Others panic at the reduced varieties of seeds available to farmers today, worrying that this loss of genetic diversity may cripple the ability to respond quickly to new outbreaks of pests and diseases. Some simply resent multinational companies and fear that a diminishing number of corporations, each becoming larger and more powerful, are becoming a force stronger than government or any other influence in our society. Still others fear the crops themselves, worrying about their safety for human health and their potential impact on our already diminished natural environments.

Whatever their benefits, risks, and impacts, genetically modified crops would not have been possible without the foundation and infrastructure established by the seed industry and particularly by Roswell Garst. He probably would support GM crops wholeheartedly if he were still alive, but to build on the commercial seed business that he inspired required another innovative trend that was just beginning when he died, the growth of molecular biology. It was the intersection of the seed industry and advances in molecular biology in the early 1980s that instigated the biotechnology revolution.

In the Heat of the Day

Research Triangle Park in North Carolina is aptly named. It is parklike, with most of its land planted with trees. The geometry is also prominent, with the 7,000-acre park located at the center of a 50-mile-long triangle formed by the three university towns of Chapel Hill, Durham, and Raleigh. And there is no question about the research part of the title; virtually all of its 50,000 employees work for companies whose primary function is inquiry.

Today's research park is a peculiar and contemporary concept, an evolutionary development from the previous generation's industrial park. North Carolina's version is a swath of thickly forested terrain that obscures buildings containing over 18 million square feet of laboratory and think-tank space that houses 140 corporations. But in contrast to most stretches of parkland, eight-lane highways thread their way through the rolling and picturesque countryside between the three feeder cities and the research triangle, with morning and evening traffic jams typical of those in any major U.S. city. Research Triangle Park also does not fit the stereotype of an industrial park, where isolated rural sites traditionally have been used because they hide from public view the spewing smokestacks and toxic effluents that are the by-products of many industries.

The products of the Research Triangle are comparatively clean, not discharging any obvious pollution. Rather, the biotechnology, software, instrumentation, pharmaceutical, and telecommunications companies in the park are creating new information, exploring the edges of what is known today in order to make the breakthroughs that will lead to tomorrow's corporate fortunes.

Eighteen of the companies in Research Triangle Park are biotechnology firms developing genetically modified crops and other bioengineered products. While there may not be visible pollutants flowing from these knowledge factories, their research is considered by some to result in the most toxic of all industrial effluents, genetic pollution. Viewed in that light, the charming country environment of the research park becomes more menacing, providing isolation and security for highly competitive multinational biotechnology companies that are experimenting with the very essence of life itself.

This simplistic and sinister interpretation of contemporary industrial research may contain elements of truth, but it obscures one of the most significant trends of our time, the merger between the cultures of enterprises driven by profit and scientists motivated by curiosity. Research Triangle Park exists because of this new partnership between entrepreneurs and researchers, a collaboration that is changing the nature of science and yielding a cascade of products whose potential benefits are enormous but whose potential for harm is as yet unknown.

What is most difficult to grasp about today's biotechnology companies is their need to balance the secrecy of industry with the strong traditions of openness in basic scientific research. Scientists who work for the biotechnology industry lead complex and sometimes conflict-ridden professional lives. Their jobs require them to focus on the industrial mandate to produce new products, and their public statements must be cleared by head offices to make sure they are consistent with the corporate objective of persuading the public that genetically engineered organisms are desirable and safe.

Yet most biotechnology scientists view public relations restrictions as a minor irritant, and pursue their science with the same sense of wonder and adventure as their counterparts in academia. They thrive on the opportunity to make landmark discoveries, and see the resultant products as profound contributions to both scientific inquiry and the public good. Scientists in the world of corporate biotechnology remain at heart curious and idealistic, even in a realm governed by the realities of corporate spin.

At the base of the biotechnology industry's evolutionary tree are the recombinant DNA technologies developed in the early 1970s. Countless companies sprouted from that root, with many branches formed from complicated corporate break-ups and mergers, but all expressing the common ancestral trait of fundamental scientific inquiry. The recombinant DNA controversies of that era anticipated today's arguments about the products of what was then a revolutionary technology, and articulated many of the issues and opinions that have come to dominate contemporary debate about genetically modified organisms and, more broadly, the role of science in society.

Recombinant DNA (rDNA) technology involves removing DNA from an organism in one species and recombining it with the genes of an organism from another species. This process is common in nature, especially among viruses and bacteria. After infecting a cell, a virus genome takes over and redirects the cell to

produce progeny virus particles. Some viruses are capable of inserting their genome into the host's and later commanding the
host to reproduce the virus's genes. Bacteria also can be highly
fluid in transferring genetic material, and many are promiscuous
in exchanging genes with bacteria of their own or other species.
What was unprecedented about rDNA research was the developing ability of scientists to target and exchange specific genes. Not
only was it now possible to move genes between more primitive
organisms like viruses and bacteria, but it was possible to transfer
genetic material between plants and animals.

Scientists quickly realized both the potential benefits and the
unknown hazards of rDNA research. Their excitement was palpable in the publications that began to appear in the 1970s, albeit
couched in the subdued language typical of scientific journals.
Maxine Singer and Dieter Soll, of the U.S. National Institutes of
Health and Yale University respectively, wrote in a 1973 letter in
Science: "These experiments offer exciting and interesting potential both for advancing knowledge of fundamental biological
processes and for alleviation of human health problems." In a
1975 *Science* article, Paul Berg of Stanford University and his colleagues wrote, "Impressive scientific achievements have already
been made in this field, and these techniques have a remarkable
potential for furthering our understanding of fundamental bio-
chemical processes."

Simultaneously, however, the same leading scientists who were
excited about the research realized the potential risks. The dangers of rDNA became a public issue following the 1973 Gordon
Conference on Nucleic Acids, one of a prestigious series of long-
standing summer meetings held annually in relaxed settings at
which scientists present and discuss preliminary but exciting
research. The 1973 Singer and Soll letter, calling on the U.S. National Academies (Academy of Science, Academy of Engineering, Institute of Medicine, and the National Research Council) to

study the risks, represented the concerns of a large majority of the participants at that meeting, and it succinctly summarized the potential dangers: "Certain such hybrid molecules may prove hazardous to laboratory workers and to the public. Although no hazard has yet been established, prudence suggests that the potential hazard be seriously considered . . . The conferees suggested that the Academies establish a study committee to consider this problem and to recommend specific actions or guidelines should that seem appropriate."

That letter led to the establishment of a series of committees, and eventually to reports of their findings, with the key recommendations developed during a 1975 Asilomar Conference held in California and chaired by Paul Berg. The outcome of the 1973 and 1975 meetings was a proposal to put the brakes on many experiments until hazards could be better assessed, and the introduction of voluntary guidelines to safely contain research material in the laboratory. These voluntary restrictions were closely observed by virtually all in rDNA work, and they soon became mandatory government guidelines regulated through the National Institutes of Health.

A number of potential hazards were identified. One risk was that viruses and bacteria used as vectors to carry genes from one organism into another might escape from the laboratory into the outside world, causing virulent super-diseases. Another potential biohazard was the transfer of genes between humans or animals, or from one animal species to another, that might carry unrecognized viruses or defective genes which could propagate into major threats to human or animal health. Finally, the greatest risk was identified as the possibility of transferring into other organisms DNA from highly pathogenic organisms or genes that produce toxic products.

The levels of containment proposed to minimize these theoretical risks matched the perceived biohazard potential for each

experiment. Minimally risky experiments required only com-
monsense laboratory practices, such as no eating in the lab, wearing lab coats, and disinfecting contaminated materials. Low-risk
experiments required limited access by laboratory personnel,
fume hoods, and the use of only safe viral or bacterial vectors to
carry and insert genes into the cells of their new hosts. Higher-
risk experiments could only be conducted under the following
conditions: the vectors had to be genetically disabled, so that they
could not survive outside of a test tube; the laboratories had to be
completely isolated from the outside world by reverse air flow
and decontamination chambers; and anything leaving the lab had
to be thoroughly disinfected, including the researchers themselves. The riskiest research with pathogenic organisms or toxic
genes was prohibited.

The scientists felt that the safeguards were thorough, appropriate, and had a high probability of successfully containing what
they viewed as remote hazards. Given their assessment that most
experiments were safe, the scientists believed that they had exhibited remarkable precaution in imposing safeguards even when
the potential for danger was remote.

The initial public response surprised many scientists, a lot of
whom were not prepared to deal with the subjective reasoning
that is more typical of public debate than of scientific discourse.
Objections focused on both ethical and safety issues, using the
now familiar tactics of legal challenges, political lobbying, inflammatory press releases, and street theater to mobilize public
opinion and influence regulatory policy. These perspectives and
tactics set the theme and tone for contemporary debates about
biotechnology.

George Wald, a Nobel Laureate from Harvard University, was
one of the few scientists opposed to recombinant DNA research.
He was particularly eloquent in outlining the ethical issues. In
1979 he wrote, "Recombinant DNA technology faces our society

with problems unprecedented not only in the history of science but of life on the Earth. It places in human hands the capacity to redesign living organisms, the products of some three billion years of evolution. This is the transcendent issue, so basic, so vast in its implications and possible consequences that no one is as yet ready to deal with it . . . I fear for the future of science as we have known it, for human kind, for life on the Earth."

The more commonly articulated perspective from the scientific community, that scientists had taken the potential hazards into account and developed safe containment guidelines, initially was not reassuring to a skittish public. The antirecombinant movement focused more on the lack of knowledge about rare events than on the reassurances from scientists that they could predict and contain the risks. Even the biosafety protocols mandated for all rDNA research by the National Institutes of Health's regulatory personnel began with a frank statement about risk: "The hazards may be guessed at, speculated about, or voted upon, but they can not be known absolutely." The language of recombinant science also frightened the public, with terms like vectors, phages, and bacterial hosts evoking doomsday scenarios.

However, as time passed, the reasoned voices and open debate among scientists calmed the public's concerns, and recombinant DNA research ceased to be a burning issue by the early 1980s. The guidelines worked well, and remain in place more or less unchanged today. Also, the technology itself proved less dangerous in experimental practice than its critics had feared. Indeed, by 1980 most experiments considered risky and controversial in the mid-1970s were downgraded to the status of minimal risk, with little public protest, since extensive research had revealed no problems. Almost all rDNA research today is conducted with few restrictions.

Another factor in the shift of public opinion about recombinant DNA was that the techniques used proved more illuminating about basic science than even rDNA's strongest proponents had hoped, and produced significant medical advances more quickly than anyone had anticipated. In 1978, an early biotechnology company, Genentech, announced the first medically important rDNA creation, a bacterium that produced insulin. In 1980, scientists from the University of Zurich and Harvard University who had founded the private company Biogen reported the cloning of bacteria that produced the antiviral compound interferon. The biotechnology industry had been born, and the race was on to profit from medical and agricultural products created by rDNA techniques.

Recombinant DNA for medical research began to be perceived as socially positive, and the public shifted its focus from the risks of rDNA to its benefits. Media headlines like *Newsweek*'s "DNA's New Miracles" and *Life*'s "Miraculous Prospects of Gene Splicing" reflected the growing public confidence in and support for the new technology.

Nevertheless, objections continued to be voiced, but the target of protest shifted from the safety of the research to the nature of the products resulting from rDNA technologies and the monopolistic practices of the multinational companies that developed and marketed them. The core of the debate today remains the same, ethics and safety, but the emphasis has changed from basic to industrial science, especially work involving agricultural applications such as genetically modified crops.

This change in the profile of protest followed the movement of scientists out of university and government laboratories and into industry. The earlier curiosity-driven recombinant DNA research conducted at universities has morphed into industrial laboratories focused on commercial products. The scientists have followed the money, and in their own version of recombination have mixed the excitement of scientific inquiry with the imperative of the private sector to turn a profit. A new type of biologist has been created, an entrepreneurial scientist working for a multinational company with access to staggering amounts of funding and resources.

Syngenta and Aventis are two of the conglomerated agricultural biotechnology companies whose laboratories are located in Research Triangle Park. I visited their facilities over two unusually warm days in November 2000. Both are housed in isolated buildings, architecturally indistinct from each other, set back from the park's highways, and even their reception areas provided quick insights into the nature of today's research-industrial complex.

Aventis had just undergone a corporate restructuring, and both their reception zone and the area beyond the locked security doors were in the midst of extensive renovations because the laboratory's research priorities were changing owing to the recent merger. At the reception desk were two greeters, one an outgoing, friendly woman with a pleasant southern accent radiating welcome and openness. The second was a scowling guard who seemed to have picked up an attitude from watching too many movies about southern sheriffs, and whose body language radiated security-consciousness and secrecy.

Syngenta was more difficult to find, because I was looking for a company called Novartis. However, the building in which Novartis was supposed to be located had a sign at the entrance identifying it as the home of Syngenta. The receptionist inside casually mentioned that yes, the company's name had been Novartis yesterday, but it had merged with another firm and today its name was Syngenta.

I had come to talk with some of the scientists creating genetically modified crops. The breed of corporate research scientist working for Syngenta, Aventis, and the numerous other corporations that have entered the GM crop sweepstakes emerged during the heady times following the development of recombinant DNA technology. The discoveries of the 1970s spawned a new industrial entity, small companies established by leading university molecular biologists and fueled with venture capital. By 1978 many of these spin-off companies had ballooned into industrial powerhouses, and the four leading firms of Cetus, Genentech, Biogen, and Genex had stock valued at over $500 million in combined value within half a decade.

Their explosive growth attracted the attention of larger, established corporations that wanted a piece of the GM pie. Investment by multinational companies like Hoffman-LaRoche, Pfizer, Upjohn, Eli Lilly, Standard Oil, Monsanto, and Dow Chemical resulted in complex ownership arrangements between these companies, their university founders, and conventional industry. In addition, industry began investing heavily in research, providing enormous grants to university departments previously dependent on trickles of government funding and offering equipment-starved professors work at industry facilities where they would have everything necessary to conduct their research.

The molecular biologist as entrepreneur had been born, and the impact of this phenomenon percolated from the university superstar researchers down through their students. The first effect was felt in medicine, but the prospect of genetically engineering crops soon attracted agricultural scientists wanting to follow their molecular biologist predecessors into the world of the biotechnology mogul.

Scientists remaining in academia have not been completely sanguine about the movement of their colleagues into industry. While admitting the economic benefits that have emerged from biotechnology, academics often condemn their industrial counterparts for selling out to corporate interests. Part of this dynamic is historical, since recombinant DNA became an issue just after the Vietnam War, when industry's reputation at universities was less than stellar. Sheldon Krimsky, then acting director of the Urban, Social, and Environmental Policy Program at Tufts University, squared off in a 1980 debate published in *Nature* with David Baltimore, then at MIT and a leading advocate of industrial biotechnology: "In the 1960s, many scientists (including yourself) expressed moral outrage in learning of the war-related research carried on secretly at many universities . . . Just as war-related academic research compromised a generation of scientists, we must anticipate a similar demise in scientific integrity when corporate funds have an undue influence over academic research. We need independent scientists willing to speak out. The direction of research influenced by industry may not be the direction that is in the public interest."

Concerns about the chilling effect of corporate influence on free scientific expression have continued to be expressed, even by scholars whose research has been crucial to developing new biotechnology products and who have amassed personal fortunes through genetic engineering. Arthur Kornberg, Nobel laureate and biomedical professor at Stanford University, weighed in with a November 2000 essay in the *Toronto Globe and Mail:* "Companies are not in business to do research and acquire knowledge for its own sake. Rather, they are in research to turn a profit. They possess neither the mandate nor the tradition to advance scholarship. Biotechnology companies must, instead, prove their profitability in the ebb and flow of financial markets and focus on short-term goals. We cannot let the money-changers dominate our temples of science."

Industry researchers don't view themselves quite that harshly. Rather, they consider GM crops a positive force, entailing few risks, for economic gain, agricultural production, human nutrition, and environmental protection. Perhaps naively but certainly sincerely, they don't understand what all the fuss is about.

I talked with Dean Bushey and Alison Chalmers at Aventis, and the next day Rich Lotstein at Syngenta, all obviously bright, university-educated scientists who had chosen to move into industry. They exhibited a casual but elegant style, somewhere between that of the rumpled academic and the suit-and-tie businessman. Their crisp corporate appearance would not have been out of place in a photo shoot for a J. Crew or a Lands' End catalogue.

Bushey and Chalmers work for the new Genomics Research wing of Aventis, charged with finding transferable genes with the potential for insertion into crops. Bushey is a synthetic chemist who came to Aventis from Union Carbide after retooling his skills to focus on biotechnology research. Chalmers is British; she received her Ph.D. in insect physiology and worked on insecticide development for American Cyanamid before coming to Aventis. Lotstein was Vice President for Administration and Public Affairs at Novartis, having moved from his earlier position in crop development research into management, but his future role in the new Syngenta had not been defined when I visited.

I was immediately struck by the same combination of openness and secrecy that I had sensed at the Aventis reception desk. It was clear throughout our discussions that there were limits on where our conversation could go. All three were open in discussing their excitement about the science they were conducting and their pride in their company's products, but any details about current research were strictly off limits. Their sensitivity about being interviewed did not stem merely from the need to protect proprietary industrial secrets; they were apprehensive about engaging in any debate with outsiders.

Nevertheless, as we continued talking and our kinship as scientists became apparent, their enthusiasm and sense of mission began to emerge from behind the corporate veil. All three preferred the business world to academia because to them the cutting edge of biotechnology research had moved from universities to industrial laboratories. They were no longer subject to the demands placed on university scientists to teach students and publish results, requiring the protracted process of writing and submitting papers, undergoing peer review, making revisions, and checking page proofs, a process that can impose a gap of up to two years between experiments and publication. For industry scientists, an exciting new finding can move from raw data to a commercial product in that time, in the process providing quicker gratification, a substantially enhanced impact, and considerably higher salaries.

They also have a staggering level of funding, equipment, and facilities available compared to their counterparts in academia. I toured both Aventis and Syngenta laboratories, and at each was taken past room after room filled with expensive state-of-the-art equipment, including gas chromatographs, mass spectrometers, and DNA sequencers capable of performing the myriad technical tasks needed to conduct contemporary research in molecular biology.

Moreover, they have acres of greenhouses devoted to growing and testing new genetically engineered crops prior to field trials. Lotstein took me on a virtual world tour through Syngenta's greenhouse mini-farms, first into an Iowa-like corn field, next to a hot and humid rice paddy such as can be found in the Philippines, then to a stand of winter wheat such as would be grown on the steppes of central Asia, and on through canola, soybean, and tobacco fields. These facilities and the well-paid scientists who work in them are supported by huge annual research budgets, $550 and $760 million respectively at Aventis and Syngenta.

But it is not just these scientists' liberation from the confining world of academic publishing and the resources industry puts at their disposal that drives their enthusiasm for genetically modified crops. They are genuinely proud of their role in creating a new agricultural technology that in their view has immense potential to improve human welfare. Bushey told me that "eventually the genetically modified organisms are going to win, because it's such a powerful, such a beneficial technology. I'm hoping that we'll be sitting around with our children and discussing how weird it was that people didn't want this technology. The technology is incredibly powerful to make things better." Chalmers agreed, saying that "it would be an awful shame to walk away from the science."

Lotstein was the most passionate about defending the socially positive mission of industrial scientists to improve the lot of humankind, and challenged the perception of environmental and health risks: "We think the benefits far outweigh the risks. I view biotechnology as a tool, and how can you say a tool is bad if it is applied in careful fashion, well regulated, and provides great benefits to society? I still can't find any instance where plant biotechnology has been used for negative purposes . . . I haven't heard any compelling arguments as to why nutritionally enhanced rice or yield boosts are bad for people."

He went on to tell me what one woman said at a meeting he attended where GM crops were being criticized, a speech that became a personal touchstone reminding him why industrial research is important and relevant: "I think about Florence from Nigeria who got up at a meeting where there was a lot of posturing; she talked about African women in her country during sweet potato growing season, largely women who spent eight to ten hours a day bending over picking weeds. Until you've been in that field in the heat of the day, you have no business telling us whether to grow a herbicide-tolerant plant."

⬛ The research may be personally rewarding, and the scientists themselves thrilled with the resources available to them, but life in industry has perils that their tenured and secure university counterparts do not encounter. The biotechnology industry is a highly volatile one, replete with mergers, restructuring, and sudden shifts in research strategies and corporate philosophy. Bushey, Chalmers, and Lotstein had each experienced yet another corporate merger shortly before my visit, Lotstein just the day before, and they were struggling to determine how they would fit into the new business entities that were emerging.

Still, the companies that are creating and marketing genetically engineered crops have retained a common corporate feeling through the various mergers and spin-offs, an identity derived from a lineage that can be traced back to the pesticide and chemical industries that developed during the Industrial Revolution. Both Syngenta and Aventis have stellar pedigrees from this world of industrial chemistry, but each has developed its own business strategy to move into the promising markets for GM crops.

The merger in November 2000 that resulted in Syngenta brought together two already multinational agricultural behemoths, Novartis and Zeneca, each bringing its own proud genealogy to the new mix. Novartis had resulted from a complex history that began with three chemical companies founded in the eighteenth and nineteenth centuries, Geigy, Sandoz, and Ciba. In 1970, Ciba and Geigy merged to form Ciba-Geigy, later renamed Ciba, while over the next decade Sandoz acquired a number of chemical companies. The Ciba and Sandoz empires merged in 1996 to form Novartis. Zeneca's history began with the 1926 merger of two chemical companies, Brunner and Mond, to form Imperial Chemical Industries (ICI); thereafter a bewildering series of acquisitions and sales led to the formation in 1999 of Astra-Zeneca, which fused with the agribusiness component of Novartis into Syngenta the following year.

One or the other of the two parent companies of Syngenta was involved with a number of agriculture's most famous and in-

famous pesticides, such as the chlorinated insecticide DDT, an array of triazine-based herbicides including atrazine, the alpha-
napthylacetic acid–derived herbicide 2,4-D, and the group of
herbicides that include diquat and paraquat. Most of these are either banned or rarely used today, but the profits generated from
these first- and second-generation pesticides fueled the entry of
Syngenta's ancestral companies into the biotechnology field.

Ciba, Sandoz, and ICI knew about genetically engineered crops
from the early pioneering work done by Monsanto (see Chapter
10), and to catch up they began acquiring seed-producing firms as
outlets for the bioengineered seeds they intended to create. By
the mid-1990s, their descendant companies Novartis and Zeneca
had become tightly interwoven hybrid firms merging chemical,
biotechnology, and plant-breeding components into multifaceted
agricultural empires. Their earliest GM product came out through
Ciba in 1995, under the trade name Maximizer with Knock/Out,
and was the first Bt corn sold in the United States.

Aventis resulted from a similar array of corporate buying and
selling, with a mix of chemical and seed ancestry, but today is following a business strategy different from Syngenta's. Its two ancestral lines, the German-based Hoechst Aktiengesellschaft and
the French Rhône-Poulenc, merged in 1999 to form Aventis.
Hoechst had begun manufacturing synthetic dyes from coal in
1863, later had moved into fertilizers and pharmaceuticals, and by
the 1930s had become IG Farbenindustries AG, whose corporate
reputation was badly tarnished before and during World War II
by its close relationship with the Nazi regime and the use of slave
laborers in its factories, particularly those supplied from the concentration camp Auschwitz. Rhône-Poulenc had a better war
record, but because of its production of pesticides it had a check-
ered environmental history.

The Aventis merger initially seemed confusing because the

new company had two disparate components: Aventis Pharma, which made prescription drugs, vaccines, therapeutic proteins, and medical diagnostic products, and Aventis Agriculture, which engaged in plant biotechnology, seed production, and pesticide production. However, in November 2000 Aventis announced plans to divest itself of its agricultural divisions and focus on pharmaceuticals. This shift was not quite what it appeared, though, because selling its agricultural components did not mean it was getting out of the crop-protection business.

The corporate restructuring created complex interactions between Aventis and about thirty other companies that it owned in part or had agreements with to produce the genetically engineered products invented by Aventis itself. Dean Bushey showed me a chart he had created that traced the relationships between the new Aventis family of firms. He pointed to a company in which one division of the Aventis team had purchased a minority interest in another company already owned by Aventis, and joked, "We're buying ourselves up!"

The company's new strategy was to use the biotechnology group headed by Chalmers and Bushey to invent new GM crops, but turn the field testing, regulatory procedures, and marketing over to other companies that it owned in part, or else license its inventions to independent firms. Through this scheme, Aventis retained the patentable and lucrative intellectual property that GM crops represent, but separated itself from the costly production and politically sensitive public relations involved in bringing genetically modified crops to a commercial level. Some of this convoluted corporate restructuring may have been driven by the StarLink controversy, in which corn from Aventis's Bt variety StarLink made its way into food products although it had been licensed by the U.S. Environmental Protection Agency (EPA) solely for use as animal feed (see Chapter 5). This incident cost

Aventis $100 million to buy back its GM corn from farmers, the loss of the StarLink registration that allowed Aventis to sell the corn seed, and a considerable amount of public good will.

The other biotechnology corporate behemoths are Dow Chemical, DuPont, and Monsanto, each with a similar history, beginning as chemical manufacturers and moving to transgenic crops through mergers, acquisitions, and spin-offs. Monsanto, for example, began in St. Louis in 1901 to make saccharin, and its product lines gradually diversified to include plastics, food additives, nylon, ammonia, and pesticides before the company went on a buying spree to acquire biotechnology-related industries in the 1990s. Monsanto itself was purchased a few years ago by an even bigger company, the pharmaceutical corporation Pharmacia, but Pharmacia soon reorganized so that all of its agricultural products are now grouped together independently as a wholly owned subsidiary under the Monsanto corporate umbrella.

Although these dizzying strategies of buying up, merging, and reorganizing companies may be confusing to corporate historians, they certainly have been profitable. Syngenta earned a net income of $564 million in the year 2000, from sales of $4.9 billion, and had over 20,000 employees worldwide. Aventis employed 95,000 people in 120 countries that year, with annual sales of $19 billion and $700 million in profits. The same year Monsanto had 14,000 employees generating $149 million in profits on $5.5 billion in revenue.

The melding of corporate practices and scientific research has resulted in products that are viewed by industry scientists and corporate spinners alike as being beneficial for agriculture and environmentally safe. Although this fusion of the research and business communities in contemporary biotechnology firms has been financially successful, it nevertheless remains an uneasy alliance between two different cultures. The scientists are curious, open-minded, and communicative by nature and training, but secrecy, privacy, and image control are still the operating mode in biotechnology board rooms and public relations offices.

The corporate line about the science has an insincere feel of manipulative public relations about it rather than the more trustworthy objectivity we used to associate with the university scientific community. The public's mistrust is not surprising given the slick tone that industry has used in selling GM crops.

The spin that industry wants to put on biotechnology was brought home to me in the first few minutes of my visit to Syngenta. Rich Lotstein hustled me into a meeting room equipped with a VCR and a television set, and before we talked he wanted me to watch a tape about Syngenta that had been carefully designed to announce the new merger and project the desired corporate image.

The tape smoothly segued between images of biotechnology laboratories, pastoral farm scenes, and up-scale yuppie supermarkets. Idealized views of life in developing countries were prominent, with laughing children playing in villages, and market scenes showing happy customers exchanging money with prosperous-looking and equally cheerful farmers, presumably for genetically modified produce. The background music was hypnotic, and the voice-over powerfully compelling: "In a diverse world, a changing world, a quality-driven world, with increasing pressure on farmland, have you thought about how we can feed eight billion people and be environmentally responsible? We have . . . We're here to meet these challenges."

Syngenta, of course, is not unique among biotechnology companies in trying to connect a rural family-farm atmosphere with prosperity fueled by invention. The large biotechnology firms have joined together in various lobby groups to pursue their mutually held public relations objectives. Typical of their publicity efforts is a newspaper advertisement put out by the Council for Biotechnology Information, whose motto, printed in green, is "Good Ideas Are Growing." The ad depicts a rugged middle-aged farmer dressed in clean denim clothes and a carefully weathered waxed cotton coat, leaning against an equally weathered wooden picket fence, with the headline "Biotechnology is helping him protect the land and preserve his family's heritage," and a quote from him saying, "I'm raising a better soybean crop that helps me conserve the topsoil, keep my land productive, and help this farm support future generations of my family."

The text goes on to emphasize that because of biotechnology corn farmers use fewer chemicals, developing countries can better feed their growing populations, and in the future farmers will grow more nutritious food. Then, the info-advertisement brings in the irrelevant issue of health benefits from biomedical research, in a subtle attempt to associate GM crops with the public's more positive attitudes toward medical bioengineering: "Biotechnology is also enhancing lives in other ways, helping to create more effective treatments for diseases such as leukemia and diabetes."

The spin campaign comes across as manipulative and self-serving, but I heard the same message delivered with sincerity and passion from the industry scientists who are creating these new technologies. Industry messages lack credibility because of their overly slick delivery, but the inventors submerged beneath the public relations hype actually do believe that GM crops are safe, economically advantageous, healthy for consumers, environmentally positive in their potential to reduce pesticide use, and an important tool for developing countries to use in improving their food production and bring their farmers out of abject poverty.

Corporate public relations has an agenda; industry needs the public as consumers of genetically engineered food and also as supporters of relaxed regulatory rules that encourage the commercialization of GM crops. But the industry approach to consumers has been heavy-handed in promoting that agenda, leaving the underlying impression that the biotechnology industry is hiding something. Ironically, these companies have suppressed the voice of their own scientists, who can speak with the aura of truthfulness that is lacking in lobbying and advertising, and who might more eloquently and convincingly make their point.

Both Dean Bushey and Alison Chalmers at Aventis had been hesitant to meet with me, concerned about how our interview might be viewed by the management. As we became more comfortable with each other, Bushey opened up and told me that "when you called up, the question was should we talk to this guy or not. You want to do it, but you don't want to put yourself in a position where you're going to cause a problem you're going to have to justify sometime down the road. You don't get anything out of it, and you could lose a lot." Chalmers talked about the company's edicts as well as her own reticence: "If a newspaper reporter comes up to you, you were told don't talk to them . . . I have opinions which may not be those of the company, and I am reluctant to speak them. I'm an industry person now, like it or not, and I either say nothing or say what the company likes. I think industry scientists are still objective, willing to argue with anyone on the science, but we're not equipped to argue the politics." Rich Lotstein, even as a vice president in Syngenta, had to clear my visit through the company's public relations officers, who carefully scrutinized my background before approving our interview.

My forays into industry left me ambivalent about the complex scientific-corporate conglomerate that has evolved in the biotechnology business. This town-and-gown marriage has been one of the most dominant influences on human endeavors in the last fifty years, with its products yielding innumerable benefits and generating enormous wealth for inventors and investors alike. It also has raised important questions about risks, and we have yet to reach a societal consensus concerning the consequences of biotechnology for human and environmental health.

The debate about genetically modified crops in particular has been hampered by the unusual history and corporate identities of the companies inventing and marketing these products. Today's multinational agricultural biotechnology companies are a hybrid combining the traditional pesticide empire with its emphasis on benefit over risk; the descendants of university-based recombinant DNA researchers, who stress fundamental and curiosity-driven research; and the corporate media and public relations personnel, with their determination to maintain the party line in public discourse.

The merger of these three independent lines has created contemporary biotechnology corporations that are excellent inventors of products but are deeply mistrusted by the public because of their glib dismissal of consumer concerns about risk. Scientists who believe that genetically engineered products are ultimately in the public's best interest are attracted to these companies by the immense resources available for research, the opportunity to participate in exciting discoveries, and the high salaries offered. But by joining these companies, they have lost some of their independence. Their employers demand that they protect proprietary inventions and make public statements that support the companies' objectives, rather than engage in frank discussions about risks and benefits. As a result, citizens have not heard the kind of deep and wide-ranging discussions about GM crops that characterized the consideration of recombinant DNA technology in the 1970s.

The intense debate about the safety and ethics of recombinant DNA research was informative, revealing the power of honest dialogue between scientists and the public to resolve issues about the negative consequences of scientific progress. Today's industry scientists are no less thoughtful than their predecessors from those debates; they are equally passionate about their science; and they are deeply committed to proceeding with the products they invent only if the benefits outweigh the risks by a large margin. Yet their opinions are heard only by the few interviewers who make it past the corporate bulwarks.

Perhaps our benchmark question about contemporary science and industry should be to consider what would have happened if Harvard University's critic of recombinant DNA research, George Wald, and others like him, had worked for industry. Could they have objected to the speed at which biotechnology was proceeding, and if so would appropriate safety guidelines have been implemented, and would the public's welfare have been enhanced rather than diminished?

People sitting in the room are rearranging deck chairs on the *Titanic*. I have no stake in what the Environmental Protection Agency decides except that we're going to sue them.

Doreen Stabinsky, Greenpeace Science Advisor for the Genetic Engineering Campaign, interviewed at an open meeting of the Scientific Advisory Panel on Risks and Benefits of Bt Plant Pesticides, Environmental Protection Agency, 19 October 2000

The Regulators

Washington, D.C., is an inspiring city, replete with soaring monuments to past American heroes, iconic buildings like the White House whose profiles evoke patriotic fervor, and archives and museums filled with timeless treasures of art, history, and technology. Limousines crowd the streets, conveying diplomats, senators, executives, and presidents to weighty meetings held in opulent hotel suites and formal ceremonial offices.

Crystal City, Virginia, is just across the Potomac River from the monuments, the museums, and the consequential decision making, but it is not an inspiring city. Visitors are more likely to arrive by the Metro rapid transit line than by limousine, and to meet in functional rather than opulent surroundings. Nevertheless, it is here that much of the real business of government takes place, within nineteen nondescript office buildings and ten business-standard hotels, connected by a labyrinth of subterranean passageways linking underground clusters of fast-food restaurants, dry cleaners, newsstands, secretarial services, and quick-copy stores.

Part of the Environmental Protection Agency, whose mandate is to protect human health and safeguard the natural environment, is located in the Crystal City complex. There the staff of the EPA's Office of Pesticide Programs interpret, implement, and enforce the higher-level decisions made across the river at the EPA headquarters in Washington.

This is the realm of the regulators, the dedicated but overworked mid-level bureaucrats who translate the stewardship responsibilities of the EPA as defined by Congress into practical policies. Their decisions mediate between the apple pie and motherhood objectives of health and environmental safety and the reality of the innumerable interest and lobby groups that come to Crystal City hoping to influence how the EPA goes about its business.

Genetically modified crops fall under the jurisdiction of the EPA's pesticide watchdogs, since some of these crops deliver the same insect-killing substances found in sprayed insecticides. The insecticidal bacterium *Bacillus thuringiensis* has been of particular interest to the EPA, and ironically so, because sprayed Bt is considered to be the safest insecticide ever invented, environmentally friendly enough to be used even by organic farmers.

Bt engineered into genetically modified crops poses issues that the sprayed formulations do not, since Bt crops provide continuous exposure to the insect-killing bacterial toxins whereas the sprayed version is active for only hours to a day or two. Environmentalists, farmers, and the EPA are concerned that season-long exposure of pest insects to Bt toxins will induce pest resistance, rendering this flagship biological pesticide useless and negating forty years of its safe and effective use. The prolonged presence of Bt in the environment also might hurt non-target organisms, especially butterflies and moths, and putting Bt into a crop rather than spraying it onto crops might harm human health by concentrating its toxins in food.

Bt crops have been the focus of many regulatory meetings, and undoubtedly will be the focus of many more. One such meeting took place from 18 to 20 October 2000 in Crystal City, and it exemplifies how pesticide regulation works. The meeting's purpose was to allow input to three scientific advisory panels composed of an overlapping membership of about thirty university and government scientists, and to provide a forum for the panels to deliberate in public before making recommendations to the EPA about the continued use of Bt crops.

Attending the three-day meeting were stressed EPA officials, pesticide industry representatives, politically astute farmers, angry environmentalists, data-hungry scientists, and the ever-hovering reporters, hoping that someone would drop a bombshell in an unguarded moment that could headline the evening news. The gathering of these diverse pressure groups in one hotel meeting room was revealing, indicative of the underlying issues of who gets to have a say, how regulators work, and the pathways by which technological change is implemented and regulated.

The unstated agenda for the EPA was to impose a higher level of regulatory oversight than had previously been required for conventional pesticides. The decisions made before and flowing from this meeting are fundamentally altering the philosophy underlying pesticide registration, but industry at one end of the spectrum, the environmental movement at the other, and science as interpreted by both sides continue to tug the regulators in all directions.

These three days in Crystal City highlighted both the strengths and the weaknesses of the regulatory system. There is much at stake, and the EPA is striving to strike the right balance between benefit and risk. Whether or not the agency gets it right with Bt crops is important, because these regulatory decisions will determine the future of biotechnology in agriculture.

███ We have come to assume that the strong hand of government will protect us from risks posed by new technologies, but the government's regulatory arm is a relatively new phenomenon. Pesticide regulation burst on the scene in 1962, when the publication of Rachel Carson's book *Silent Spring* stimulated a tidal wave of protest and lawsuits, culminating in the creation of the Environmental Protection Agency in 1970. The EPA moved quickly to ban DDT, the poster-child of the new synthetic chemical pesticides, and then grew to regulate virtually every aspect of modern technology that produces pollutants as by-products. Today, the EPA has a $7-billion budget and over 18,000 employees dedicated to protecting us and the environment from our own worst excesses.

The need to regulate biotechnology emerged rapidly following the recombinant DNA discoveries in the 1960s and early 1970s. At first, all biotechnology regulation came from the National Institutes of Health (NIH), but other government agencies began to get involved when the biology moved from basic science performed in laboratories to commercial applications. The specific need to regulate agricultural biotechnology became apparent in 1983, with a proposal from industry to field-test "Ice Minus," a group of genetically altered crops with genes producing antifreeze proteins that protect plants from frost damage. This proposed field trial was met with cries of anguish from environmental groups, and stimulated a series of legal challenges that forced the federal government to expand its regulatory involvement with GM crops.

By 1986, the Office of Science and Technology Policy in the

White House had brokered an interagency agreement that defined lead agencies for regulating various aspects of biotechnology in agriculture. The NIH retained its authority to review the process by which a GM crop is produced, the Food and Drug Administration (FDA) was given the authority to regulate food itself, the U.S. Department of Agriculture (USDA) received the rights to ensure that new crop plants developed through biotechnology did not themselves become pests, and the EPA was given authority over genetically engineered organisms designed to produce pest-killing substances.

The existence of several jurisdictions means that a complex, multistep process is required before a genetically modified crop receives full approval. The first stage rests with the NIH, which determines if the methods used to produce the crop are safe. This step has become a rubber-stamp stage, because the recombinant DNA techniques needed to develop a new GM crop have been used extensively by scientists and industry since the 1970s without incident, and are deemed generically safe by the NIH.

The next registration stages involve the safety of the product itself, the GM crop. The USDA assesses whether the existing research facilities are adequate to conduct the research, then approves field trials. After field-testing is completed, the company creating the new seed submits a voluminous, data-heavy report to the USDA to prove that the crop poses no risk to U.S. agriculture and can be deregulated and grown without further USDA oversight.

The role of the FDA is to determine the crop's safety as food, and its approval process focuses on what is called "substantial equivalence," as well as on allergenicity. The standard of substantial equivalence requires industry to demonstrate only that the proteins, carbohydrates, fats, and/or oils from a GM crop do not differ in structure, function, or composition from substances currently found in food. Similarly, passing the allergenicity test requires only proving the lack of known allergens. Both standards have been criticized as too weak by GM opponents, since adhering to them requires testing only for known substances, not for new substances or unknown interactions between previously identified food constituents (see Chapter 5).

The EPA's regulatory procedures overlap the roles of the other agencies and have become the most time-consuming and demanding of the standards administered by any of the government's biotechnology overseers, because the EPA has taken the strictest view of its regulatory responsibilities. The EPA grants experimental permits for any test area over ten acres in extent, and thoroughly reviews data on human, animal, and environmental safety. Its final step, which typically takes eighteen months, is a formal examination of the product covering virtually all areas of its use and impact.

The convoluted, interagency maze that developed to regulate genetically modified crops is not surprising, given the fundamental problem that pervades the regulatory process. The quandary that government regulators find themselves in is simple: protecting human and environmental health should be rooted in objective, data-based science, but in practice politics influences regulatory decisions. Many regulatory deliberations also are conducted in private, making use of proprietary and thus confidential industry-generated data. Consumers, environmental activists, and scientists have fought hard against this regulatory secrecy, aggressively and persistently pursuing Freedom of Information requests to liberate background data from the closed rooms and dark corners of government agencies.

Another endemic regulatory dilemma is the perceived need to reach a decision in a timely fashion, which is at odds with the fact that collecting sufficient data to satisfy scientific criteria is a prolonged process. Environmental issues are particularly troublesome in a regulatory timeframe, because environmental impact may take years or decades to measure while the industries argue for decisions within months so that they can proceed with commercializing products and generating sales and profits. The criteria for evaluating environmental impact also are difficult to define, and scientists struggle with objectively deciding how to measure the impact and what level of impact is acceptable in complex systems in which many processes are operating, both natural and caused by humans.

It is the lack of balance between the role of data and the role of social, economic, and political factors that scientists find particularly frustrating. It is at the regulatory junction that scientists interface most directly with nonscientists, and there that the implementation of scientific discoveries proceeds or fails. The ultimate decisions about product safety made by the regulatory gatekeepers are driven by public relations, pressure groups, and politics, although one hopes they are tempered by objective scientific analysis. That real-world mix of self-interest and secrecy is not descriptive of the way science operates, and the world of influence and lobbying is a milieu that scientists find foreign and uncomfortable.

The Crystal City scientific advisory panel meeting was particularly interesting in this broader context of regulatory dilemmas, because these panels reflect an attempt by the Environmental Protection Agency to reveal more data publicly and to make its decision-making process more transparent and science based. The regulators who were responsible for handling the hot potato of GM crops came to their task with a sense of both foreboding and opportunity. Experience with the previous generation of synthetic chemical pesticides had shown how difficult it can be to strike the appropriate balance between environmental protection and innovative technologies. Yet the issue of GM crops gave the EPA a chance to show that wiser regulatory decisions can be made publicly and with appropriate safeguards for the environment and human health.

The predicted benefits of crops engineered with Bt genes were especially seductive in their potential to decrease pest damage while simultaneously reducing the use of conventional pesticides. Bioengineered Bt crops held out the dual promise of increased agricultural productivity and financial gain for economically besieged farmers. These benefits were timely, since conventional insecticides were becoming increasingly ineffective as insects became resistant to every class of chemical that science could invent. Further, of all the pesticides, conventional synthetic insecticides inflict the most damage on non-target species, both humans and beneficial organisms in the environment. Sprayed formulations of Bt were widely viewed as safe alternative insecticides, although they were not as effective as the growers would have liked because the efficacy of the Bt applications was short-lived in the field.

The EPA weighed the benefits and risks of Bt crops, and in the mid-1990s approved their commercial registration for corn, cotton, and potatoes, all crops that had significant insect pest problems. There was, however, a condition required for these registrations that had not been imposed on applications for approval of conventional pesticides. The regulators took the unusual step of broadly seeking counsel from outside the usual industry sources, and approached an array of independent university scientists for advice. Their input led to one of the largest regulatory experiments in the history of American agriculture, the imposition of resistance-management guidelines as a requirement for the registration of Bt crops.

The EPA's first tentative steps toward imposing the resistance-management requirement on Bt crops led it in the early 1990s to Fred Gould, a professor at North Carolina State University in Raleigh. His research on insect resistance was highly respected in the scientific community but at the time had only been published in scientific journals and presented at professional meetings, not disseminated into the mainstream of practical application or regulatory influence.

Gould's background was typical of that of many academic scientists. Born in New York City in 1949, he had excelled at college but took some time off to travel and work before going to medical school. A chance encounter at a party with a professor from Stony Brook University's Department of Ecology and Evolutionary Biology persuaded him to change direction and study natural history instead of medicine. Stony Brook's program was heavily abstract and model-focused, a quantitatively rigorous approach that he both enjoyed and excelled at.

His research interests centered on the evolution of insect resistance to pesticides, including the genetic basis of resistance, physiological mechanisms through which resistance was induced, and models predicting how long it would take for particular pests to become resistant under different pesticide-application regimes. This work might have remained known only to evolutionary biologists rather than spread throughout the practical realm of industry and farmers, but a fateful 1988 article published in the journal *BioScience* brought Gould's work to the attention of the larger community outside academia.

This now classic and prescient piece was written early in the development of GM crops, seven years before the first commercial registrations were approved by the EPA. Gould pointed out that the important potential of GM crops to reduce the use of chemical pesticides was threatened by the inherent risk of inducing pest resistance through their unregulated use: "Genetically engineered, pest-resistant cultivars could help limit the use of environmentally disruptive synthetic chemicals. As such, pest-resistant genes are an important natural resource. But if these genes are widely used in homogeneous commercial cultivars, pests may adapt to them and cause rapid loss of this useful resource. Incentives from government to stimulate public and private sector breeders to develop efficient methods for producing genetically heterogeneous crops is likely to be essential in any effort aimed at long-term conservation of pest-resistance genes . . . These new approaches will, however, require greater intellectual and capital investment."

The concept that Gould and a few other scientists such as Bruce Tabashnik and George Georghiou were promoting—to actively manage to prevent resistance—was new for EPA. The conventional agrochemical industry and farmers had failed to exercise self-control in moderating their use of sprayed synthetic chemical pesticides, and pest resistance to these chemicals had become epidemic throughout agriculture. Insecticides are a powerful selective agent, because when overused they kill all the susceptible individuals in a population but leave a small number of resistant insects to mate with each other, thereby increasing the frequency of the resistant trait in the next generation. Growers then apply higher levels of pesticides to obtain control, further selecting for even more resistant insects in subsequent generations. Moreover, many insects resistant to one pesticide are also resistant to others, so that the pesticide industry and farmers were finding themselves increasingly unable to keep up with nature's genetically based response to their chemical strategies.

One sprayed pesticide for which resistance had not developed was the bacterium *Bacillus thuringiensis,* in part because it has an unusually short period of activity after application. Bt is a common soil-dwelling bacterium that naturally produces toxic substances which disrupt the stomach membranes of some insects. The bacteria then feed on the nutrients released by the broken gut cells. Various strains of Bt produce different toxins specific to the digestive systems of certain hosts and are not effective against other insects or vertebrates, so that the impact of Bt on non-target organisms is minimal.

Bt has been the most frequently used biological control available over the last forty years, but Bt sprays are expensive, active for only hours, and not always successful at penetrating the insect gut. Biotechnologists quickly perceived that a better way to deliver the Bt pesticides would be to insert the bacterial genes producing the toxins directly into the plants on which pest insects feed. They created corn, cotton, and potato plants that secreted one of the toxins produced naturally by Bt. The toxic proteins themselves were given scientific designations such as Cry1Ac, Cry1Ab, Cry3A, and Cry9C (Cry is short for "crystalline," because the toxins come in crystallized form), but eventually sold under more evocative product names such as NewLeaf, Bollgard, Knockout, YieldGard, Nature-Gard, and StarLink.

A number of agrochemical and seed companies invested hundreds of millions of dollars in the race to produce Bt crops, including Monsanto, Novartis Seeds (now Syngenta), Mycogen, DeKalb Genetics (now part of Monsanto), and AgrEvo/PGS (now Aventis). The first registrations to use Bt crops commercially were approved by late 1995. Bt potatoes did not prove to be necessary for potato pest management, since other methods were working well and economically, and thus they were not widely adopted. Bt corn and Bt cotton, however, swept through the agricultural sectors, so that by 2000 close to 30 percent of both crops were genetically modified with Bt genes.

Gould and others who studied pesticides had come to understand many of the factors that predisposed insects to rapidly evolve resistance. Two aspects of Bt crops were of particular concern to the resistance-savvy scientists. First, the early versions of these crops contained relatively low concentrations of the insect-killing toxins, so that insects with a mild tolerance to an insecticide would survive and contribute to resistance in the next generation. The second problem was that pests were exposed to the growing Bt plants for an entire season, which allowed a number of generations of resistant insects to mate and fix the resistant traits into the pest population. Thus the developing Bt crops had excellent potential to rapidly induce textbook cases of insect resistance.

Laboratory experiments with pests exposed to Bt confirmed the potential for resistance, and led to the development of theoretical models which predicted the likelihood that resistance would develop under various agricultural scenarios. The models, untested in the field, suggested two methods that could dramatically reduce the probability that resistance would develop. One hypothetical technique was to increase the dose of the Cry toxins in plants to at least twenty-five times the minimum necessary for control. This high-dose strategy was designed to kill even the most resistant insects. A second idea was to create refuges, areas in or adjacent to fields that would be planted with non-Bt crop varieties. These refuges could be border strips of non-GM varieties planted next to Bt crops, plots of conventional and Bt seed mixed and planted together in a mosaic, or blocks of non-Bt crops planted as rows within Bt fields. Whatever their design, the refuges were meant to be areas where unselected pests could thrive, providing a susceptible population that would interbreed with Bt-resistant individuals and thereby prevent the resistant trait from becoming dominant.

The models developed by Gould and others were theory, not fact, and were untested in field situations. The methods they recommended were difficult for industry and farmers to accept, partly because they were new and unproven but also because they had costs associated with them. The high dose was of concern to industry because it required additional expense to develop crops that secreted a large amount of toxin, and because consumers and environmentalists would be even more concerned about these higher doses of toxin being contained in food plants. The refuge strategy was equally problematic for farmers, since it meant planting 5 to 50 percent of their fields with susceptible varieties that would be attacked by pests, reducing yields in those sectors and possibly negating yield gains from the Bt crops. Also, just physically planting any crop was a difficult enough task with one variety, and farmers feared that requiring them to plant different varieties in various patterns would be an unworkable complication in real-world crop management.

Nevertheless, the EPA began studying the prospects for adding the emerging resistance-management strategies to the regulatory framework for Bt crops. I talked with Gould about the EPA's initially tentative and then increasingly enthusiastic support for resistance management. We met over lunch at a small off-campus Mediterranean restaurant filled with the smells of frying falafel and the sounds of Lebanese music. Gould is clearly bright and thoughtful, relishes a good question and carefully considers his answers, not to formulate a politically correct response but because he genuinely enjoys debating agricultural and regulatory issues.

He first described the EPA's reluctance to actively espouse resistance management for conventional crops: "EPA's been pushed by people on insect resistance for pesticide management for a long time; they've responded to it and talked about it but they hadn't done anything about it other than to push industry maybe to go in that direction. I think they felt that was not their mandate."

Bt crops, however, provided a new opportunity, so EPA representatives began studying issues involving GM crops and resistance by attending meetings of the Entomological Society of America where resistance management was being discussed. Soon, resistance management "took on a life of its own. They started seeing Bt as a natural resource. I don't know if it was the organic farmer push or politics or image or what it was, but they pushed us to give them a resistance-management plan. They really got involved in it and had lots of meetings."

In 1992 the EPA began convening scientific advisory panels to solicit advice about Bt crops, with Gould as a key figure, attending or chairing the sessions. By 1995 the EPA was ready to recommend, and soon required, that Bt crops implement insect resistance management (IRM) strategies, including high doses of toxin and mandated refuges. The registrations also imposed monitoring for compliance and regular scientific reviews to determine if the theoretical IRM strategies were successfully suppressing resistance under field conditions. The October 2000 meeting in Crystal City was the final assessment of IRM and other aspects of the health and environmental safety of Bt crops before the EPA decided whether to continue their registrations, and provided an opportunity for the supporters and critics of EPA's policies on GM crops to once again have their say.

The scientific advisory panels that met over three days to review Bt crops revealed a process high-minded in theory and flawed in practice, but perhaps the best that can be done in reaching adequate regulatory decisions on highly contentious issues. Resonating through the formal meetings and the corridor conversations at the breaks was a tension between the scientists on the panels, who required extensive data before they would advise, and the EPA regulators, who had to make decisions based on the information available at the time. Also apparent were the tensions between industry and agriculture on the one side and environmental groups on the other, creating intense lobbying efforts within the constricted rules under which the advisory panels operated.

The Crystal City meeting was tightly structured in accord with the legislation defining the mandate and procedures of the EPA's scientific advisory panels. The purpose of these panels is to provide advice concerning regulatory decisions. The legislation ensures not only that the EPA seek outside expert opinion, but that the panelists the EPA chooses are independent scientists, without any financial stake in the outcome. Any person or organization can submit written comments before a public meeting. The panels then convene, hear oral submissions, and discuss the issues before submitting a written report.

Public these meetings may have been, but the room arrangement and meeting structure left no doubt that this was to be a well-controlled forum. The Bt SAP meetings were conducted in a plush, red-carpeted, lavishly chandeliered hotel ballroom. The panel and EPA representatives were seated around an enormous square table arranged so that most members of the panel had their backs or sides to the audience. A microphone was available for use during the public-input phase, with the presenters facing the panel but not the public when speaking.

Each panelist had a microphone and frequently a laptop computer, and scurrying around the table were a dozen or so EPA support staff adding to the already-high piles of paper stacked at each panelist's seat. The audience occupied most of the few hundred seats that filled the remainder of the ballroom. The symbolic message of this spatial arrangement was clear; the panel were meeting in public, and individuals could approach the dignitaries to speak, but the deliberations, dialogue, and decisions were confined to a tightly knit group.

The panels themselves consisted primarily of university and USDA scientists. The membership of the panels changed somewhat each day, depending on the issue being discussed. Each day's format was similar, with a brief introduction and summary reports presented by EPA officials, followed by strictly timed five-minute opportunities for public comment. In the concluding segment the panel members discussed among themselves specific issues about which the EPA was seeking advice, with no opportunity for input or discussion from the audience. The panelists had to put in long hours, each meeting beginning at 8:30 A.M. promptly and ending late into the evenings, and the final written reports were due a few weeks later.

The EPA organizers were tense, strained, and testy, obviously on edge about the proceedings. Their frequent private huddles in the corners and hallways throughout the meeting suggested that orchestrating the meetings required considerable concentration, and there was an undercurrent of political maneuvering throughout the proceedings. Their intensity was due in part to the many television, radio, and print journalists who were present, partly because this was the largest public turnout ever seen for an SAP meeting, but also because a lot was at stake for the EPA. The fundamental purpose of the panel was to deliver external judgment on what had become a critical element of the EPA's GM crop regulatory plan.

The meeting began with various EPA officials describing the ideal regulatory world they were striving to achieve and stressing the importance of the decisions the panels would make: "Multidisciplinary bridge between producers and basic science on campus"; "A cooperative approach for the best solutions to insecticide resistance management between producers, companies, and researchers"; and "From the huge piles of paper in front of you, you can see the enormous amounts of time and energy expended under public scrutiny on this issue of enormous importance to the nation. The eyes of the world are upon you today."

The voluminous written submissions, evidence presented from scientific research, and testimony from industry and farmers all pointed to the same conclusion. To the relief of the regulators, the theoretical plan proposed for resistance management seemed to be working in practice. After five years there were no documented instances of any pest insect having developed resistance. With cotton there were two years when one pest seemed to be vaguely moving in the direction of becoming more tolerant to the Bt variety, but then that tolerance dropped back to background levels and did not reappear. Further, industry reported more than 90 percent compliance rates with the regulations for planting the refuges, indicating that the critical goal of getting farmers to accept the resistance-management paradigms had been met.

Oddly, however, there was no feeling of euphoria or celebration in the room; instead ongoing tension was palpable throughout the panel sessions and the public presentations, and even in the hallways. There seemed to be three sources of this anxiety: the scientists themselves, who disagreed about how to continue managing against resistance; industry, which was concerned about the increasing number of regulations; and the environmentalists, whose voice was subdued throughout the proceedings but who believed that the rosy picture being painted at the meeting table was an illusion.

The scientists' concerns revolved around whether the IRM plan would continue to be effective, and whether there were enough data to refine resistance-management practices. Their models suggested that five years of success was not surprising, but also that unless stricter guidelines were implemented resistance could develop within the next five to ten years. The biggest scientific issues concerned the refuges themselves, particularly whether growers were truly compliant and whether insecticide spraying should be permitted within the refuge areas.

The reported compliance of more than 90 percent seems excellent, but the theoretical models were based on 100 percent of farmers using properly constituted refuges in which susceptible pests could thrive. Farmers don't want any pests to thrive, and providing areas in which to deliberately grow pests on crops is not an agronomic practice that intuitively appeals to farmers. Consequently, the refuges actually set out in their fields tended to occupy their least productive land, were often not irrigated, and were frequently sprayed with insecticides to reduce pest damage and increase crop yields. These practices all degraded the likelihood of success in real-world practice, reducing the number of susceptible insects and increasing the probability that resistant insects would mate with each other.

The solutions posed by the more aggressive resistance-management advocates on the SAP included doubling to quadrupling the dose of toxin in the plants and also requiring larger refuge areas, up to 40 or 50 percent larger in some Bt crop areas if farmers were permitted to spray in the refuges. These rigorous recommendations accounted for a principal source of the tense atmosphere, a persistent disagreement between resistance modelers, who saw doom approaching, and the farm community and industry, which were begrudgingly meeting the current regulatory requirements for resistance management but clearly did not want to go further. They argued that no problem had emerged after five years of practical field experience, so why should they be swayed by scientists crying wolf?

This view was strongly voiced by the cavalcade of industry and farm representatives who attended the entire three days of meetings, keenly observing the proceedings, lobbying in the hallways, and presenting their carefully worded inputs in the five-minute slots provided for the expression of public opinion. Representatives from seed companies, corn and cotton associations, and nonprofit groups friendly to biotechnology approached the panel with the same essential message, paraphrased as follows:

We support insect resistance management, and admit that it was a good idea for the EPA to encourage management against resistance with mandatory refuges and high-dose requirements.

Remember, however, that farmers must have flexibility to control pests in unprotected, non-Bt fields, and while compliance is important the regulators must take into account the problems of individual farm management. Also, industry has already invested considerable resources in creating higher-dose plants, and it would be unreasonable and uneconomic to increase the dose yet again. Producers and industry can't afford to comply with stronger regulations mandating larger refuges, reduced insecticide spraying within refuges, and higher Bt toxin doses in crops. Although we have worked together so far, we may have to withdraw funding from scientists who are developing further theoretical models if they lead to stricter regulations.

These well-articulated but blunt statements to the panels indicate the intense effect of high-pressure lobbying on the agricultural regulatory system. Regulators occupy the middle ground by mission and disposition, balancing the myriad interests and attitudes that come to bear on issues of benefit and risk. Mixed into this balancing act is the inherent caution of scientists, with their passion for models and intense appetite for ever increasing amounts of data. For scientists, there never is enough information to make an informed decision. Devotion to models sometimes becomes an end in itself, the result being an overemphasis on the role of theory when empirical information from the field is lacking.

On the other side of the regulatory divide is the enthusiasm of industry for its products, which overvalues the products' benefits and minimizes their side effects and long-term consequences. Bouncing between industry and science are the farmers themselves, whose financial bottom lines are hemorrhaging red ink and who desperately embrace new technologies in the hope of cutting a few dollars per acre off of their production costs. Industry and farmers also share a desire to be independent, preferring to be left alone by government to pursue production and profit.

There is another component to the brew, a group whose voice

has been shunted to the fringes of regulatory policy decisions concerning Bt crops and other issues involving biotechnology in agriculture. This group is composed of the critics, who in earlier debates vehemently opposed commercializing GM crops but whose perspective has become increasingly irrelevant. The gulf between regulators and the environmental movement has widened, because the regulators focus on balancing benefit and risk while the environmentalists regard any risks as unacceptable.

I came to think of them as the Gaia Sisters, three women in their forties with Ph.D. degrees in biology who had left academia to work in the environmental movement as scientific advisors and advocates. Rebecca Goldburg came to the Crystal City meeting as a representative of the Environmental Defense Fund, based in New York City, Jane Rissler spoke for the Washington-based Union of Concerned Scientists, and Doreen Stabinsky spoke for Greenpeace's Genetic Engineering Campaign.

Each presented brief public testimony during which she tried to find a way to broaden the panel's debate. Rissler was first, and began by complimenting the panels and the EPA on their homework. But she quickly moved from flattery to the heart of her fundamental concern about regulation, which was that the panel's "assessment was not dominated by good science but by industry. The EPA has produced a rosy picture of Bt crops, and caved in to industry . . . We shouldn't be pushing products ahead while waiting for good data. This research should have been done five years ago."

Stabinsky was next, sandwiched between a representative from Monsanto and a scientist hired by industry to evaluate the refuge models. She didn't bother with any pleasantries, but tried to undermine the assertion that resistance had not developed by focusing on the occasionally tolerant insects found in some years on cotton. Goldburg was the last of the three to appear before the panels, and her voice was the most reasoned and compromising. Assuming that regulated Bt crops would not go away, she chose to spend her five minutes arguing that the increased refuge areas supported by some panel members were good for farmers in the long run because they would slow down the onset of resistance.

I was intrigued by these brief blips of blunt environmental advocacy in the midst of an otherwise byzantine and nuanced process, and arranged to meet later with the three women. I met with Rissler and Goldburg first, at the office of the Union of Concerned Scientists near the White House. Their office was surprisingly corporate-looking, although informal, with a board room, computer work stations, and ergonomically comfortable chairs that could just as easily have been found in a biotechnology or e-commerce firm as in the headquarters of an environmental lobby group.

Rissler and Goldburg had been battling the acceptance and spread of GM crops for over a decade, and they both seemed tired and pessimistic. Nevertheless, they persisted in attempting to influence the outcome of what they view as foregone regulatory decisions. Rissler was particularly cynical about how predictable the outcome seemed, perhaps because after receiving her Ph.D. in plant pathology from Cornell University she worked for the EPA for some years, first as a fellow appointed by the American Association for the Advancement of Science and then as a full-time employee. Thus she knew the system as both an insider and a critic. She brought her perspective on the EPA into clear focus quickly: "The EPA assesses with a bias toward industry. They bring together scientists for advice, but the scientists favor industry. On the whole, the EPA is an enabler for industry. Agribusiness runs the pesticide program at the Environmental Protection Agency."

As we talked, I began to understand why groups like the Union of Concerned Scientists and the Environmental Defense Fund were having trouble accepting regulatory paradigms. They believe that we need fundamental changes in agriculture, and regulating what in their view are bad policies, such as the adoption of GM crops, misses that basic point. Goldburg, whose doctoral degree was in ecology, touched on that perspective when discussing GM crops: "The Bt crops are one more step on the pesticide treadmill. Both the Bt crops and the herbicide-resistant crops are part of the way we do industrial farming in the United States. They don't really change the system." Rissler followed that up by reiterating her complaints about the power of agribusinesses, saying that "they have so much clout that we can't get sustainable agriculture on the agenda."

Both women also criticized the Crystal City meeting for its failure to provide any meaningful opportunity to present points of view not based on the relatively narrow assumptions adopted by regulators. I asked Jane Rissler why she even bothered to attend: "Well, it's the only game in town. You get five minutes, and after I talk there's stunned silence. We never ever have an opportunity really to have a full discussion from our point of view of the issues because there's not anyone on the panel who will spark that or maintain that kind of discussion."

But the women did not view their decade-long quest to influence decisions about genetically modified crops as a total failure. Although they would prefer that no GM crops be planted, they do believe that they have affected the conditions under which these crops can be grown commercially. As Rebecca Goldburg put it, "I really think there would have been no regulations without environmental organizations. Without the environmental community we wouldn't have resistance management." Rissler agreed: "I think we had an effect with regulators in slowing down this technology through stronger regulations than there would have been without us."

I talked with Doreen Stabinsky from Greenpeace the next day, when we both needed a break from the incessant day-long drone of regulatory language. She received her Ph.D. in genetics from the University of California at Davis, but then almost immediately began working as an analyst and advocate for environmental organizations. Her position is simpler than Goldburg's and Rissler's. Greenpeace wants to ban all GM crops, period, and so the issue of resistance management is irrelevant to its agenda. Stabinsky attended the Crystal City meeting not with expectations of moderating the regulatory decisions but to gather information for an upcoming lawsuit.

Stabinsky's stance is clear: "We definitely have a radical position on this, but we're a radical organization. One of the biggest problems of genetic engineering in agriculture is that it's put into a vision of conventional agriculture, and used to maintain a system that concentrates production into just a few hands. These forces are problematic for all people who are eating. My biggest problem with genetic engineering is social and political—it's a symptom of manipulating nature for our own profit and using nature as a disposal facility and not recognizing its intrinsic value. I'm outraged by that."

Rather than trying to influence the regulatory system, Greenpeace, the Sierra Club, the Center for Food Safety, and sixty other organizations and individuals issued a notice of intent to sue the EPA, announced in a press release put out during the Crystal City meeting. Their approach was clever, designed to fault the EPA for not following government regulatory policies rather than to fight the agency on the science of resistance management. The pending lawsuit was based on section 7 of the Endangered Species Act, which states that each federal agency "shall insure that any action authorized, funded, or carried out by such agency . . . is not likely to jeopardize the continued existence of any endangered species."

There is as yet no firm evidence that Bt corn is contributing to the decline of any endangered species, but the actual impact of Bt corn is not the lawsuit's focus. Greenpeace and its fellow organizations allege that the EPA has not prepared the legally required biological assessment or formally consulted with and obtained official biological opinions from the Departments of Interior and Commerce. The intent-to-sue letter was clear about the consequences. If the EPA fails to comply with the "statutory requirements concerning the impacts on threatened and endangered species from genetically engineered Bt plants, then the listed parties will have no alternative but to seek relief in federal court."

The time I spent in that Crystal City hotel was among the most draining meeting experiences I have ever undergone. I left the meeting late Friday afternoon in a daze, and took the Metro into Washington, where I could decompress by wandering around on the spectacular urban parkland of Independence Mall. As I regained my composure, the source of my fatigue and disorientation gradually clarified. I had just spent three days being manipulated, with every hallway conversation and interview designed to shift my opinion, just as each submission to the panel members was intended to change their perception of the issues. The meeting room was like a pressure cooker, with interest groups scheming to influence what went into the pot and how it was mixed.

What was lacking was dialogue, an honest and straightforward exchange of views. The regulatory process does not allow true discussion. Orchestrated and controlled, it is fueled by coercion and pressure. Lobbyists from all sides present their carefully crafted positions in an attempt to push their perceived self-interest to the forefront rather than to participate in genuine dialogue in order to reach a well-balanced conclusion. In this atmosphere, the radical differences between industry, consumer, and environmental positions on genetically modified crops become starkly apparent, and the polarized stances this issue has generated become abundantly obvious.

Yet it is important not to lose sight of what the EPA has achieved within the highly politicized boundaries of regulation. The registration of Bt crops was inevitable, given the benefits they provide, their lack of demonstrated as opposed to possible side effects, and the strong influence of industry on the Environmental Protection Agency. In that context, the imposition of resistance management as a registration requirement was nothing short of remarkable, especially given the failure of the agricultural community to adopt resistance management for chemical pesticides even after generations of farmers overused them.

There may be something warped about a process in which pure pressure from interest groups is so dominant, and it is discouraging to realize that true discourse about issues with vast economic and environmental implications cannot be achieved within government regulatory systems. Nevertheless, the decisions made at such regulatory proceedings may be the best we can achieve when dealing with highly charged issues such as GM crops, and the progressive outcome of the SAP meetings on Bt crops suggests that in spite of its nature the regulatory process may be working, even if laboriously and imperfectly.

Fred Gould's comment on the regulatory dilemma points up both the limits and the potential of this important process: "How much power does a single regulator have in determining the course of things? I know they struggle with each other, and industry is always at their door, and so are the environmentalists— there's an incredible amount of pressure. Why are they taking resistance management so seriously? The group at EPA personally has a stake in that. They're not doing this as bureaucrats. They'll feel good about their lives if they can get this thing to work."

> One conflict is between the rights of some people to exploit the environment in pursuit of a livelihood, as against the rights of others who want to preserve the environment as an amenity.
>
> **Nuffield Council on Bioethics,** *Genetically Modified Crops: The Ethical and Social Issues,* 1999

Of Butterflies and Weeds

Picture the midwestern United States in summer, a bountiful and boundless vista of cornfields broken only by the occasional cluster of farm buildings and the checkerboard pattern of gravel roads and lanes. Focus, now, on the edges, the in-between places and roadside borders where a few scraggly weeds have survived the annual onslaught of herbicides and mowing. It is at these edges that the battle for the genetic integrity of our environment is being fought, and there that the last refuge of prairie wildlife is encountering its latest foe.

Genetically modified crops have many potential side effects, but none has aroused the public ire as much as their potential for environmental damage. The operative word is "potential," because a clear consensus has yet to emerge about the long-term impact of GM crops on their natural counterparts in the field. The rhetoric has been extreme, the scientific data scanty, but that is beginning to change as slow-moving granting agencies have gradually begun to fund research to help resolve this most important of issues.

The areas of concern fall into three major clusters. First, environmentalists worry about wind-blown pollen from GM crops settling on leaves of both crop plants and nearby roadside vegetation, killing caterpillars and other organisms that feed on plants. Second, the same pollen carried by wind or bees might cross-pollinate closely related plant species, or the seeds from GM crops could escape and become feral, dramatically intruding on nature. And third, GM crops could harm non-target beneficial organisms, especially parasites and predators that feed on pest insects or bees that provide the important service of pollinating crop and weed species alike.

Today, the data are beginning to emerge, and the picture they provide does not support the extreme views of either the pro-GM or the anti-GM forces. Rather, the impact of GM crops on the environment surrounding them is proving to be conditional, dependent on the crop, circumstance, and how the issue is approached by the scientists studying it. These incoming data are providing some basis for judgment and decision, but because the scientific studies are showing a mixed bag of environmental spillover from GM crops, we find ourselves in the classic bind presented by most new technologies. How do we predict and balance benefit and risk, and does a compromise position at the center satisfy those at the extreme ends of the spectrum?

The real or imagined side effects of GM crops also pose a deeper question. We already have transformed nature, terraforming most of the arable land on our planet to a degree never before induced by a single species. Given our already vast impact, will the effects of GM crops even be noticed, or should GM crops be our environmental line-in-the-sand?

Environmental concerns are tinderbox issues, smoldering for years before a press release about a threatened species provides the additional fuel and a provocative research finding provides the spark that ignites a full-blown public fire. For the impact of GM crops on the environment, monarch butterflies were the threatened species and a report from Cornell University about the potentially detrimental effects of GM corn pollen on monarch caterpillars was the spark.

Monarch butterflies have always captured the public imagination, being one of the few insect species to evoke admiration for their beauty and grace. North Americans are attracted to monarchs partly because of their summer presence throughout much of the continent, where they can be seen feeding on milkweed plants as caterpillars and lofting lazily on summer breezes as adults. The caterpillar larvae feed on the leaves of the hundred or so milkweed species in North America, ingesting the plants' toxins, which then make the larvae and adults poisonous to vertebrate predators. The adults also are unusual, and almost unique as insects, in that they migrate like birds for thousands of miles each fall to winter in the southwestern United States and central Mexico before returning north in the spring to repeat their life cycle.

Their unusual feeding and migration habits have been the subject of a considerable body of scientific research, focused on understanding the milkweed-monarch interaction, and on investigating the complex of good-tasting butterflies that mimic monarchs in order to fool predators into thinking that they, too, are poisonous. In addition, the ability of adult monarch butterflies to find the same wintering sites used by their deceased parents, which they themselves have never visited, has led to a wealth of studies explaining their orientation mechanisms and the evolutionary advantages of this strange insect life style.

Their few overwintering areas are fragile, located in places threatened by urban development in California or logging in Mexico. Pacific Grove, California, is a good example of both the splendor and the fragility of these sites. Pacific Grove calls itself the "Butterfly Town, U.S.A." because of its small grove of Monterey pine and cypress trees that host a wintering population of monarchs each year. The butterflies roost in this pastoral grove because it is shady and sheltered, with regular winter fogs providing moisture and keeping the butterflies cool to preserve their energy reserves.

To visit this or other wintering sites is a transcendent experience, peaceful and other-worldly, with hundreds of thousands of colorful monarchs festooning the trees like cloaks, and the occasional errant butterfly flying eerily through the mist to return to its clustering cohorts. Yet the grove is small, only a few acres in area, and is surrounded by the city, adjacent to a strip of motels with names like the Butterfly Grove Inn and the Monarch Resort, which house tourists who come to admire this small outpost of nature. But however splendid, this and other wintering sites have been preserved only through concerted efforts by butterfly aficionados.

Monarch butterflies are also well known to the public because they are commonly studied in American and Canadian science courses. The development of courses using monarchs was spearheaded by the Monarch Watch program based at the University of Kansas. This program, funded by the U.S. National Science Foundation, has sent out monarch butterfly larvae and educational kits to hundreds of thousands of students since its inception in 1991, and involved students in research projects to track the migration of these beautiful insects.

Monarch butterflies have been of interest to conservationists for many years, because in spite of their seemingly large population of about 100 million individuals they are threatened by human activities. Urban sprawl, pesticides, and agricultural development have reduced the monarchs' habitat in their northern summering grounds. Their only food plants, the milkweeds, are considered noxious weeds and frequently are attacked with herbicides. Monarch overwintering sites in the south remain vulnerable, protected for the moment but continually imperiled by expanding human populations and human activity in the surrounding regions.

Thus public interest in and awareness of monarch butterflies is not new, and it is not surprising that some of the first studies on the effects of GM crops on non-target organisms focused on these unusual insects. But it is not only monarchs that could be harmed by bioengineering; many species of butterflies and moths are potentially susceptible to the current generation of insect-resistant crops. These plants have been engineered to carry a gene transferred from *Bacillus thuringiensis* that is designed to kill pest insects such as the larvae of the European corn borer, but can also be toxic to other caterpillars.

Corn is the crop of greatest concern to butterfly admirers and environmentalists, because the copious amounts of pollen produced by corn could be blown by the wind to and settle on nearby vegetation that is eaten by monarchs and other non-pest caterpillars. If so, the toxic protein expressed by this gene in the corn pollen might devastate these innocent insect bystanders. Such an event could have serious ecological repercussions, because butterfly and moth larvae are important food sources for a web of bird and other vertebrate predators.

One of the first studies on the side effects of Bt corn was published by John Losey and his colleagues from Cornell University as a short note in the prestigious journal *Nature*. It served as the catalyst for a cascading series of studies, claims, and counterclaims about the effects of GM crops on butterflies. Their study was done entirely in the laboratory, and consisted of feeding monarch caterpillars milkweed leaves dusted with either conventional corn pollen or pollen from genetically modified Bt corn. The results were dramatic. After four days, butterfly survival was virtually 100 percent among caterpillars that ate conventional pollen, but only 56 percent among those that ate GM pollen. The surviving larvae also ate less and were significantly smaller when fed GM pollen.

The authors were both cautious and alarmist in interpreting their results, using language such as "potentially" to recognize the preliminary nature of their data, but also pointing to the "profound implications" implied by their findings. They followed traditional scientific etiquette in calling for more studies, pointing out that it was "imperative that we gather the data necessary to evaluate the risks associated with this new agrotechnology." Nevertheless, they concluded their paper by suggesting that Bt corn might not be the worst pest-management practice from the monarch's point of view, and that all genetically modified crops needed to be compared with conventional pesticides to determine which was the least likely to harm monarch butterflies and other non-target species.

Their caution was overlooked by the media and fellow scientists alike, but the alarm rang many bells. Pro-GM scientists quickly jumped into the fray, writing letters to *Nature* and talking with the media to point out the shortcomings of Losey's study. In particular, critics noted that the caterpillars were fed only one dose of pollen, an apparently high one, and that this dose might have no relevance to the amount of GM pollen blown onto milkweeds in the field. One particularly upset member of the biotechnology community told Losey, "People like you should be on the endangered species list." Anti-GM groups also leaped in with a predictable flurry of we-told-you-so press releases and Web site postings.

Industry responded quickly, and in a cunning move funded a consortium of scientists to conduct field studies. Their objective was to determine how much and how far corn pollen spreads onto milkweeds, and whether the amount of pollen in the field was any real threat to monarchs. Funding for seventeen studies was dispersed from the Agricultural Biotechnology Stewardship Working Group, a subset of the American Crop Protection Association, a consortium of biotechnology and pesticide companies. The group wisely directed the funding to a range of independent university and government scientists rather than to corporate laboratories, deflecting any criticism that the results might be biased.

Calling on independent scientists may appear to have been a high-risk move, but in fact industry had an inkling of how the results might turn out from unpublished field studies already conducted in the midwestern corn belt. These and new field studies conducted in Maryland, Iowa, and Ontario found that there is very little corn pollen deposited on milkweeds outside of corn fields, and even at field edges the quantity of corn pollen on milkweeds generally is lower than the amount needed to induce effects on caterpillar growth or survival. There is almost no monarch mortality outside of corn fields, and the worst mortality found within fields has been less than 20 percent. Even this may not be particularly worrisome because adult monarchs seem to prefer laying eggs on milkweed plants that are outside of corn fields.

As these emerging data have begun to calm the initial panic about monarchs, other butterfly studies on less charismatic species also are yielding similarly soothing results. One key report on the common black swallowtail butterfly was released in June 2000, conducted by Catherine Wraight and others from the University of Illinois laboratory led by the eminent entomologist May Berenbaum. Black swallowtails might be particularly vulnerable because their host wild parsnip plants commonly are found in the narrow strips between corn fields and roads, and the caterpillars are actively feeding at the time of year when corn sheds its pollen.

The Entomology Department at the University of Illinois is a fitting institution and Berenbaum's laboratory staff an excellent research group to provide an impartial analysis of environmental risks from biotechnology. Historically, the entomologists at the university have pursued basic studies of insect biology rather than the applied studies more common at state universities. Thus they are perceived as being more independent of agricultural interests, and more likely to focus on what their data say rather than on what their biases lead them to expect.

Their study tested the effects on swallowtails of Bt corn pollen deposited on wild parsnip plants. They were looking for any relationships in the field between swallowtail mortality and the amount of pollen on leaves as well as the proximity of parsnip plants to corn. They also fed various amounts of genetically modified corn pollen to swallowtail caterpillars in the laboratory.

Generally, they found no effects of Bt corn pollen on caterpillar survival. Levels of pollen found on parsnip plants in the field did not kill larvae, and there was no connection between parsnip distance from corn fields and swallowtail death. Further, even amounts of pollen that were fifty times higher than the highest pollen density observed in the field had no effects when fed to butterflies in the laboratory. The one exception was pollen from a little-used variety of Bt corn, called event 176, which has about forty times more Bt toxin than do more commonly used varieties of Bt corn. Event 176 did kill butterflies at the highest pollen dose tested, and also had been identified as a more problematic variety in some of the monarch studies.

Studies are ongoing, but the developing picture from the monarch and swallowtail work indicates only a minor impact of GM pollen deposited on plants and eaten by butterfly caterpillars. The few effects found occur only in or within a few feet of planted corn, and at high pollen doses rarely found in field situations, and even then only with some varieties of GM corn, especially event 176.

Two steps, one involving management and the other the research sector, can be taken to further reduce even these minor spillover effects from Bt corn. First, event 176 expresses too much toxin, which is undesirable not only because of its greater potential to kill non-pest butterflies but also because these high doses will be a strong selective force favoring the development of resistance in pests. For these reasons, plants of this variety are no longer being used. Second, many GM crops are designed to express their active ingredients in some parts of plants but not others, and the Bt gene could be restricted so that it does not produce toxins in pollen.

The take-home scientific message from the butterfly work is that there may be effects from GM crops, but the demonstrated impact so far has been minimal, and appropriate management practices can reduce even these limited side effects. Future research, of course, could identify as-yet unrecognized problems, but the first blood drawn in the butterfly wars has not presented us with unresolvable damage or a much-feared apocalyptic environmental disaster.

If we want to conserve monarchs, swallowtails, and other wild butterfly species, errant GM pollen may not be the first issue on which we should be spending our time. Conventional pesticides may be considerably more damaging than GM crops. Drift of pesticides sprayed by air onto vegetation adjacent to fields is common, and probably is a more pervasive threat to non-target insect populations than wind-blown genetically modified pollen. If anything, the use of Bt corn might favor rather than reduce wild butterfly populations, and could be the more ecologically correct pest-management tool. Remarkably, no one has yet examined the impact of conventional pesticides on butterflies adjacent to crop fields, but a number of laboratories are now preparing such studies.

There is another take-home message, that balanced scientific studies indicating minor and rationally acceptable environmental impacts may not be enough to comfort either the pro- or anti-GM forces. The monarch and swallowtail studies highlight our dilemma about GM crops—and indeed all new technologies. When have sufficient studies been done to verify that GM crops are safe, and at what point should potential or demonstrated risks slow or halt biotechnological advances? Where is the equilibrium point between fact-based decision making and emotionally driven perceptions that are rooted in bias rather than data?

Losey's critics from the biotechnology industry might be surprised to learn that he is cautiously pro-GM. He is a young scientist, still pre-tenure at Cornell, and is articulate in expressing his perspective, balanced in his views, and refreshingly enthusiastic about his work. He talked to me about being caught between the environmental rock and the industrial hard place: "You're trying to be fair and objective and you're not really liked by either side. Misinformation was happening on both sides of the issue. Neither side's response was valid, and neither reflected what we said. All we're saying is that we found this phenomenon in the lab and we should study this further. I don't think Bt corn is having such an effect that we should ban it, but I'm not ready to say that it's completely butterfly safe. I would say proceed with caution."

I also spoke with May Berenbaum about how difficult it can be to provide a balanced scientific perspective on emotionally charged biotechnology issues that will satisfy the environmentalists. Her background lends credibility to both her science and her influence in the public arena. She is a member of the National Academy of Sciences and a member or chair of innumerable high-level national committees involved with insects, pest management, and science teaching. And she is also a prolific and humorous writer, and has written several popular books that have provided her with more recognition and public credibility than the average entomologist. She has an engaging and computer-quick mind, wide-ranging interests, and, like many entomologists, she has a quirky side. She organizes and hosts the world's largest insect fear film festival, attended annually by a thousand or so devotees of entomological horror flicks.

Berenbaum is full of energy, but even her elevated energy levels become hyper-agitated when she talks about the reactions of anti-GM environmentalists to the monarch and swallowtail studies. One called her swallowtail research an "abuse of science" because it did not find the hoped-for impact, while others demanded to know which corporate entity had bought her. For Berenbaum, the experience has "been a major source of stress and aggravation from unexpected sources. If anything, I'm one of the tree huggers. My experience with environmentally oriented groups is they are desperate to pick holes. Is this such an important issue that it requires extraordinary proof? When do you stop? If you're unwilling to believe it, nothing is good enough. Image is everything. If it hadn't been the monarch, if it had been a little moth without a common name, would anyone have cared?"

In our contemporary agriculture-intensive world, genes may be the ultimate pollutant. We already have dramatically modified our environment by moving genetic material around the globe, and by deliberately and accidentally introducing crops and weeds. Even so, there has been an unprecedented level of public panic induced by the specter of bioengineered genes jumping from GM crops into weeds, or GM crops themselves becoming weedy. The facts surrounding the likelihood of gene flow are elusive, but a growing databank has begun to define the probabilities and impacts of genetic movement from crop to nature. The developing picture shows outcomes similar to the butterfly case: Genes will jump, and there will be impact. Our decisions about if and how to proceed will depend on how we balance agricultural productivity and human-directed genetic infiltration of the environment against the integrity of nature.

Basically, there are two ways for genetically modified genes to infiltrate nonagricultural terrain. First, pollen carrying a gene that has been introduced into a crop could be carried by wind or bees from the bioengineered crop to pollinate a closely related weed. Plant species hybridize more easily than animals, and a gene for herbicide or insect resistance could become ensconced in a wild plant, with unknown consequences. Second, the crop plant itself might go feral and become a super-weed because of its resistance to pests. If so, the bioengineered gene might confer an ecological advantage to the recombined crop, enhancing its survivability outside of managed fields and also interfering with our ability to manage weeds in and away from agricultural habitats.

The question of whether genes actually could jump to wild relatives was one of the earliest issues that scientists began to address, and they did so well before genetically modified crops became commercially available. This issue is most problematic for GM crops in the geographic regions where their conventional predecessors originally evolved. It is in those locales that crop plants are most likely to have close feral cousins that could most easily be pollinated with pollen carrying novel genes. An amazing 98 percent of crops in North America originated elsewhere, including corn, wheat, rice, rape, potatoes, sugar beets, and soybeans, and so there are fewer but still some weedy relatives to be worried about there. Concern in Europe has been more intense, since most European crops originated locally and have many wild relatives.

The answer to whether genes can jump from plant to plant, even for conventional crops, is a clear "yes." Hybridization between related plant species is a common phenomenon in plants, which exhibit more promiscuous gene flow between species than is found in animals. There are hundreds of examples of genes from domesticated crops jumping to feral weeds, with somewhere between one quarter and one third of all crops known to have transferred genes to their wild relatives. More significantly, twelve of the world's thirteen most important food crops have sent genes into wild plants through pollen transfer.

The success of genes transferred from domesticated into feral plants varies considerably. Most commonly, the impact of gene flow is to reduce the fitness of feral plants in subsequent generations, so that the domesticated-wild hybrids die out quickly. However, there are notable exceptions. For example, the wild radish originated as a hybrid between the cultivated radish and a related, introduced weed species, and has become an important weed in California. Johnson grass, one of the most troublesome weeds world wide, is a hybrid between cultivated and wild sorghum species.

Sugar beets provide a good example of how fluidly genes can move between weeds and crops. Cultivated sugar beets have been selected to flower every second year for management purposes, but wild sugar beets flower annually because doing so increases their feral fitness. In France, the wild gene for early flowering spread into a sugar beet nursery, probably by pollen movement, and from there was dispersed to farms throughout France. The resulting crop created considerable problems for farmers, but also increased weed problems off the farm as seeds from the cultivated annually flowering beets spread to noncultivated areas.

These examples suggest that it would be prudent to assume that some gene movement from GM crops to weeds and back will occur, and the resultant management costs and environmental impacts should be recognized as potentially negative effects of GM crops. Plant scientists have been actively looking into the transfer and establishment of bioengineered genes in wild plants, and oilseed rape in Europe has become one of the lead organisms for this research.

There are a number of weed species related to rape to which genes carrying herbicide or insect resistance might travel, including hoary mustard, charlock, and wild turnip, cabbage, and radish. The first question researchers asked about gene flow involved the distance that pollen might move from crop fields. Wind-carried rape pollen can be blown almost a mile from rape fields, and bees can carry the pollen even longer distances. Thus there would be an opportunity for GM oilseed rape pollen to encounter the flowers of feral weeds over a wide area surrounding planted fields.

If GM pollen is carried to flowers of closely related weed species, will it pollinate successfully, will the seed be viable, and will the GM trait be expressed? The answer to each question is a qualified "yes." Pollen from oilseed rape can produce viable herbicide- or insect-resistant hybrids with many weeds. This phenomenon was first reported in the mid-1990s, when Thomas Mikkelsen and his colleagues from the Risø National Laboratory in Denmark grew transgenic rape in small plots interspersed with wild turnip. Fertile, recombined turnip seeds resulted that expressed the herbicide-tolerant rape characteristic when grown in the next turnip generation.

The qualification is in the frequency that this occurs, and in its significance. These between-species sexual recombination events are relatively rare, but since they can and do occur, subsequent generations could pass on transgenic traits if they confer improved fitness on the progeny of feral weeds. Genes for either herbicide or insect resistance transferred through oilseed rape pollen have not been found to provide any unusual selective advantage to closely related feral weeds in the absence of herbicides or insect attack, but neither do they have any negative fitness cost for the plant. The resistant genes can be maintained over at least five generations if left to their own devices, and herbicide applications or insect outbreaks favor the resistant trait.

A related issue is the weediness of GM crops themselves. That is, will crops go feral, and will their life outside of farmers' fields have any negative consequences? Even conventional crops spread and grow outside planted fields, and can persist for many years as weeds. Oilseed rape, for example, has been found growing wild in regions of Scotland ten years after it was last cultivated. Escaped volunteer crops generally have been considered to be amenable to management where they need to be controlled, because herbicide treatments can eliminate them on roadsides, beside hydrological rights-of-way, and in farm fields when crops are being rotated.

The potential for herbicide-resistant crops to become feral weeds has aroused more concern than the same potential for conventional crops, for the obvious reason that they would be resistant to chemical management. There already are 216 herbicide-resistant weed species found in forty-five countries around the world. These naturally occurring weeds evolved protective chemical defenses after many decades of heavy herbicide use, and maintain the trait because of continued spraying. Indeed, one argument favoring herbicide-resistant GM crops is that their use would reduce herbicide spraying on cropland and lower the selective pressure favoring the evolution of resistance in non-target weeds.

The flip side of this argument is that a genetically modified herbicide-resistant crop gone weedy would create an instant herbicide-resistance problem, without waiting for evolution to do its work. These volunteer weeds might persist and become dominant, serving as a permanent source of weed problems and having unknown effects on plant and animal biodiversity. Biotechnologists counter-argue that seeds from herbicide-resistant plants would rarely end up outside fields, and if they did so could easily be managed by spraying an herbicide different from the one to which they are resistant.

Sounds reasonable, except that in the spring of 2000 a Canadian farmer from Sexsmith, Alberta, discovered something that was supposed to be virtually impossible, triple-resistant canola seed. He had planted three different herbicide-resistant GM canola crops over a two-year period, in adjacent fields, and the following spring wanted to "clean" his fields of any remaining canola before planting wheat. However, he discovered that the volunteer canola sprouting in and around his fields could not be eliminated by spraying any of the commonly used herbicides—Roundup, Liberty, or Pursuit. Resistance to each of these weed-killing pesticides had been bioengineered into one of the canola varieties he had planted, and apparently the genes conferring resistance had been transferred between plants sequentially to produce triple-resistant hybrids.

This phenomenon of gene stacking is probably rare, and can be avoided by not planting different herbicide-resistant varieties in close proximity to each other. Further, triple-resistant volunteer canola can still be controlled by spraying a fourth herbicide, 2,4-D, because so far resistance to 2,4-D has not been incorporated into a GM crop. Yet it remains troublesome to contemplate the ease and speed with which the herbicide-resistant genes combined to create a super-resistant weed with few management options.

However, one long-term study from Britain suggests that GM crops will not necessarily survive for prolonged periods outside farmers' fields. M. J. Crawley and his colleagues from Imperial College in London tested herbicide-tolerant oilseed rape, corn, and sugar beets as well as two types of insect-resistant GM potatoes, and found that escaped GM varieties had poor survival rates in natural habitats. Most of the GM plants had gone extinct within two years and all had disappeared within four, which indicates that for those varieties in that habitat transgenic crops going feral will not be an issue. This finding may have broader applicability for most GM crops, since they have been selected to thrive with the fertilizer and mineral inputs typically used in conventional farming, inputs not available in otherwise competitive wild settings.

The best strategy for farmers to use in managing weeds remains debatable. Will GM crops reduce herbicide spraying and thereby lower the selective pressure on feral weeds to evolve resistance, or will the resistant gene itself spread from crop to weeds, and resistant crops go feral? The outcome of this large-scale natural experiment will be crucial both for the success of weed-management strategies and for maintenance or establishment of habitat diversity in and around agricultural regions, yet the predicted outcomes depend on the perspective of the predictor.

Environmentalists focus on the worst-case scenarios of intensive gene flow and the escape of crops into the wild, while farmers and the biotechnology industry remain confident that we can stay one step ahead of any problems by rotating crops and using diverse herbicides on and off the farm. In the end, this issue may revolve more around philosophy than around science: Are we willing to accept increasingly intensive management in order to utilize the advantages of GM crops, or has our impact on nature reached a point where it's time to call a halt to further intervention?

Another focus of environmental concern about GM crops is that they may have adverse effects on the food chain by influencing non-target beneficial organisms such as predators, parasites, and pollinating bees. It has proven particularly difficult to obtain data about this potentially important impact, partly because any food chain studies are inherently complex but also because industry and the government agencies that regulate GM crops have not been forthcoming about the research they have conducted.

This and other information about environmental impacts often has been deemed proprietary and can be notoriously difficult to extract from either government or industry sources. Some independent university and nonregulatory government laboratories have conducted studies in an effort to counter this excessive secrecy, and also because they sense a potentially fascinating research area. Fortunately they have not yet revealed serious negative effects of GM crops on non-target, beneficial species, but publicly reported studies are still too few and incomplete.

Bees could be particularly vulnerable to GM crops because the nectar and pollen produced by plants are the sole food sources for bees, and crops today make up a substantial part of the diets of both wild and managed bee species. Nectar is the principal sugar and carbohydrate source for bees, but nectar from GM crops is not problematic for bees because it contains virtually no proteins. Of more concern is pollen, which is the only protein source for bees and in many cases expresses the same bioengineered proteins that are found elsewhere in genetically modified plants.

The limited amount of publicly accessible knowledge about GM pollen and bee health comes primarily from one source, that of Minh-Hà Pham-Delègue, who works for the French government research branch INRA (Institut National de la Recherche Agronomique) in a laboratory just outside of Paris. Her research group has investigated the effects of pollen from some GM oilseed rape and soybean varieties on adult honey bees. Both crops are important sources of pollen for managed and wild honey bee colonies, produce bioengineered proteins that could potentially harm bees, and have pollen that expresses the same transgenic proteins found in the green parts of the plants.

Pham-Delègue's group studied a particular gene transferred to both soybeans and rape that produces a protein called chitinase designed to provide resistance to fungal diseases. Chitinase interferes with the production of chitin, a substance found in the cell walls of fungi that also is an important component of the digestive system and external skeleton of bees. They also examined other modified proteins from GM soybeans that inhibit the activity of insect digestive enzymes, and inhibitors from GM rape that interfere with other insect enzymatic functions. These inhibitors are effective against insect pests, but also could interfere with honey bees if consumed in pollen.

Taken together, their studies are comforting in that they indicate none of the GM pollens kills adult bees outright, and it is only at unrealistically high doses that any long-term effects on adult bee survival have been found. However, GM pollens can induce more subtle behavioral changes in adult bees, particularly in their ability to learn, which could be potentially harmful because bees leave the nest to forage and must remember their way home. Caution is advisable when interpreting these results, since there is no information from field studies to indicate whether this learning disability occurs in the field or if so whether it is important. Further, their research so far has been restricted to only one bee species, the honey bee, and only to adult bees. Field work and larval studies are the next obvious and essential steps before we clear GM crops of any accusations that they harm beneficial bee species, and all we can say so far is that a few varieties of two GM crops do not kill adult honey bees directly.

Another area of concern about GM crops is that they may harm parasites and predators that control insect pests. Many of these natural biological control agents are insects themselves, such as parasitic wasps that lay eggs on or inside host insects, or predacious beetles, bugs, or lacewings that consume prey directly. These useful insects could be vulnerable to modified proteins consumed by their hosts or prey. If so, an undesirable side effect of bioengineered crops would be the reduction or elimination of beneficial organisms, and an increased dependence on pesticides or genetically modified crops to manage insect pests.

The potential for GM crops to interfere with the pest and predator food chain has been demonstrated in the laboratory, but has not proven significant in field studies. For example, lacewings feeding on European corn borer larvae in the lab showed almost twice the mortality when the corn borer had eaten Bt corn instead of nonmodified corn varieties. Similarly, lab-grown potato aphids that had fed on GM potatoes modified with anti-aphid lectin proteins in turn caused reduced fecundity and longevity when eaten by predacious ladybird beetles, although no acute toxicity was found. However, field studies have not confirmed these effects in potato or corn, and wasps parasitizing diamond-back moths feeding on Bt oilseed rape also have shown no impact of the GM crop on the parasites.

For bees and beneficial predators and parasites, the available data indicate only minor effects of genetically modified proteins moving up through the food chain. Clearly, continued and expanded research would be useful to confirm the minimal impact of GM crops in field situations, and to evaluate new varieties of bioengineered plants to ensure that they do not have harmful impacts on useful and ecologically important organisms. Continued research and vigilance would be prudent, but an ecological disaster has yet to be confirmed. Nevertheless, the comfort zone for environmentalists would be enhanced if all of the data about non-target organisms that industry is required to submit to gain regulatory approval of GM crops were unrestricted. Regulatory agencies are not doing the public a great service by releasing this information in summary form rather than with the full details required for proper scrutiny.

The emerging data concerning the environmental effects of GM crops is not neutral, but neither is it catastrophic. Bio-engineered crops may have minor effects on non-pest butterflies, moths, and bees, and perhaps occasional impacts on beneficial predators and parasites, but these effects do not appear to be serious or consistent enough to warrant banning the use of genetically modified crops in agriculture. The impact of GM crops on plant biodiversity beyond farmed fields may be more of a problem, caused by cross-pollination between crops and weeds and by GM crops going feral, but even here the impact appears to be manageable. Although new crops will require continued vigilance, and longer-term studies of current crops are desirable, the magnitude of the effects demonstrated so far appears to be no greater than that seen under the already-existing management practices used in conventional, pesticide-heavy agriculture.

It is difficult to reconcile environmental concerns about GM crops with our acceptance of the environmental impacts of conventional pesticides, or with our tolerance for radical changes in biodiversity induced by agriculture itself and by the international movement of plants and animals involved with agriculture. Even the worst-case scenarios of GM crop effects on our purportedly natural environment are turning out to be less severe than doomsayers' predicted, and GM crops may prove beneficial through their ability to reduce pesticide use.

The current environmental impacts of conventional chemical pesticides can be considerable, and the use of GM crops already is demonstrating its potential to reduce the use of chemicals in agriculture. Genetically modified crops have shown reductions in net herbicide and insecticide use in most studies to date, including analyses of herbicide-tolerant soybeans, cotton, and corn, and insect-resistant cotton and corn. In total, nine studies have indicated 5 to 10 percent reductions; one study found no change for herbicide-resistant cotton (although a second cotton study did reveal decreased herbicide use); and there have yet to be any studies that found an overall increase in pesticide use. Thus the touted benefit of GM crops in reducing applications of synthetic chemical pesticides appears to be realistic, and a clear environmental plus for biotechnology.

It also is instructive to compare the potential environmental impact of gene flow from genetically modified crops to the effects we have induced by accidentally and deliberately introducing crops and their accompanying pests. The impact on the North American environment through gene movement resulting from such introductions has been gargantuan, and it is difficult to imagine a scenario in which bioengineered crops would make modifications in our environment more radical than those already induced by the transportation of colonizing species.

David Pimentel of Cornell University has made a career out of amassing compelling statistics that illuminate just how heavy our footprint has been on the earth. He and his colleagues published a study in the January 2000 issue of *BioScience* reporting on the environmental impact of nonnative species in the United States. Fifty thousand species have been introduced to the United States, some deliberately but many accidentally. Some have been beneficial for humans, especially crops, but many have clearly been detrimental to the native environment, displacing indigenous species, overrunning ecosystems, and driving approximately four hundred resident species to the brink of extinction or beyond through competition and predation. Further, the movement of species around the globe has been irreversible, forever changing our environment into a very different biological world.

Even the worst-case scenarios for GM crops would not be more detrimental or environment-changing than these conventional invaders. Any environmental change might be undesirable, but the impact of invasive species is considerably greater than the minor to moderate ecological shifts that might be induced by the escape of genes created by agricultural bioengineering. Further, the frequency with which genes jump from GM crops to feral relatives, or GM crops themselves become weeds, and the resulting impact on the environment, both appear to be similar to what happens with conventional crops. The likelihood of genes escaping should be reducible in the future by preventing the expression of modified genes in pollen, but even if genes escape they should be controllable through the same types of management strategies currently used to keep weeds in check.

Science is telling us that the effects of GM crops on non-target organisms and their potential to become weeds through pollination or escape appear to be acceptable and manageable risks. However, opinions about the environmental spillover from biotechnology do not hinge on this science, but instead depend on our individual tolerance for management and human-induced changes to our ecosystems.

All human endeavors affect the world around us, and no activity has more environment-bending influence than agriculture, even the conventional varieties. Whether GM crops are within or outside our comfort zone in terms of their impact will depend on where we draw the line between tolerable side effects and intolerable environmental damage, and between intense management and leaving ecosystems to their own devices.

Both proponents and opponents of GM crops agree that eventually some significant ecological impact will become apparent. Supporters, including most scientists, conventional farmers, and the corporate community, generally are comfortable with the philosophy of management that accompanies contemporary agri-

culture, and believe that the gains from GM crops will be sufficiently beneficial and the ecological impact controllable enough
to justify their commercial use. Opponents focus on predictions
of eventual and uncertain ecological impact, and object to implementing biotechnology if there are any side effects that might
induce ecological change or require a management response.
These critics condemn the planting of any transgenic crops.

Conventional agriculture combines concentrated management with an acceptance of ecological side effects, and this is unlikely to change with GM crops unless some collateral damage
emerges that is considerably worse than the damage from current practices. The data collected so far suggest that this is unlikely, and that while farming with transgenic crops will have
some ecological impact, it does not yet appear to be significant
enough for opponents to block the progress of biotechnology in
farming.

It Only Moves Forward

The Vogue Theatre on Granville Street in Vancouver, British Columbia, has seen better days. Its walls are still covered with brocade, but it is frayed, and the red velvet seats with their carved wooden armrests have been worn down by generations of audiences. The location is not good, a central downtown street originally intended to be a pedestrian mall and up-scale theater district that has instead deteriorated into a tawdry area rife with sex shops and crawling with drug dealers, vacant-eyed street kids, and beggars.

The Vogue is no longer the theater of choice for big-ticket productions, but is now a venue for events with meager budgets. It is the perfect site for gatherings of protest groups rallying against the status quo, and was just right for the antigenetic engineering teach-in "Big Money, Bad $cience" held in November 2000.

The day-long event was organized primarily by the environmental lobby group Greenpeace, but had a host of other minor sponsors with radical views and angry or fringe-sounding names like Check Your Head, Raging Grannies, FarmFolk/CityFolk, Circling Dawn, and Happy Planet. The teach-in was billed as providing "balanced information about the threats posed by biotechnology and genetic engineering to the environment, public health, and sustainability."

The meeting did provide information, but it would be a considerable stretch to call it "balanced." The underlying assumption was that industry and government are duping the public into accepting genetically modified crops by suppressing and manipulating information. This assumption was implicit in the meeting's title; the science of genetically modified crops is either deficient or misused, bought and paid for by the piles of money industry has thrown into biotechnology.

Indeed, big business and big government have occasionally been caught suppressing information, and they frequently are selective in how they choose, stress, and spin their facts. However, what emerged most strongly from the "Big Money, Bad $cience" meeting was the disturbingly similar practices of the protest movement, which has been equally adept at choosing what to say and how to say it. The speakers opposing genetic engineering can be just as manipulative, buzz-oriented, exaggerated, and self-serving as their counterparts in the biotechnology industry.

The public relations sophistication of the GM protest movement has been impressive, replete with irritating tactics that have kept the industry spin doctors off balance. Although effective, the opposition teach-ins, Web sites, newsletters, and underground rumor mills can also be disorienting, spewing out a torrent of information that collectively seems bizarre and improbable. It is tempting to dismiss this protest movement as based on hearsay and innuendo rather than fact and objectivity, but although the protesters are imprecise in presenting data their message is not without substance.

To understand the appeal, impact, and validity of protest

against genetically modified organisms, it is important to recognize one key component of the anti-GM movement. The protest is not really about the science, although criticizing the science has been a key tactic. Rather, the opposition is about values and trust, and when values are the underlying agenda and trust the principal issue, the facts can become irrelevant.

The "Big Money, Bad $cience" meeting had been set up as an alternative to the industry-based Pacific Biotechnology Conference being held simultaneously a few blocks away at the upscale Hotel Vancouver. Registration for that meeting was $700, and even at that price the conference had attracted over twelve hundred participants. Admission to "Big Money, Bad $cience" was free, although a donation of from $3 to $15 was suggested, and the meeting was as much a happening as a conference.

The Vogue Theatre was almost full, with a diverse crowd of about a thousand participants. Some just wandered in from the street, having arrived on skateboards and smoked roll-your-owns outside before the program began. Others had skipped high school classes to make the scene; they arrived in bunches and were easily identifiable by their multiply-pierced bodies and spiked hair. University students also made an appearance, with overflowing backpacks and books to read at the breaks, and so did middle-aged people of the Vietnam protest generation still committed to left-wing politics. The senior citizens were represented as well, mostly gray-haired women dressed in floor-length skirts and vaguely ethnic blouses. The speakers themselves were dressed informal, most of them sporting knit vests or worn polar fleece sweaters.

Although the meeting was billed as a teach-in, the audience seemed already to have made up its collective mind, and had come not so much to learn as to cheer on the heroes of the anti-

GM movement. A legion of opposition icons were on the program, including Brewster Kneen, an activist in farm politics who wrote the book *Farmageddon;* Beth Burrows, director of the public interest group the Edmonds Institute; Maude Barlow, a political activist who chairs the nonprofit consumer advocacy group The Council of Canadians; Mae-Wan Ho, a senior research fellow at the Open University in Britain and director of the Institute of Science in Society; David Suzuki, a leading environmentalist and broadcaster; and many others.

The members of the audience had come to hear biotechnology blasted, and they were not disappointed. Speaker after speaker lambasted industry and government with rousing oratory, and the listeners were following every barb and metaphor. They gasped in agreement with horrific depictions of genetic engineering, collectively shook their heads with concern when speakers mentioned new crops being developed, and yelled "right on" when the speakers demanded action.

One recurring rhetorical theme was condemnation of the science behind biotechnology, with speakers repeatedly harping on the same myths about how scientific research had confirmed the health and environmental horrors of genetically modified crops. I found the crowd's unqualified acceptance of the strong rhetoric about the science annoying, since I had just spent months reading every scientific article I could find on the impact of GM crops and attending numerous scientific and regulatory meetings about them. What I had read and heard from the scientific community was expressed with considerably more caution, hesitation, and balance than was employed by the table-thumping orators on the stage.

Speakers claimed that monarch butterflies were in danger of extinction because of genetically modified corn, but the data suggested at worst minor population reductions, and even that conclusion was debatable. Corn and other commercialized GM crops, the speakers said, could induce fatal allergic reactions in humans, but I had not been able to find evidence that any such reaction had ever actually occurred. Bees were said to harbor bacteria that had incorporated genetically modified genes from canola pollen into their own DNA, a persistent idea that had permeated the environmental movement after it was reported in a May 2000 newspaper article that had yet to be substantiated with data. Economic studies were alluded to that proved GM crops cost farmers money, but every study I had seen found either a positive bottom line for bioengineered crops or at worst no differences in the bottom line between GM and conventional varieties.

It was tempting to dismiss the lecturers because of the loose science that was being flung around the hall, but as I continued to listen to the speakers and eavesdrop on lobby conversations, the opposition's perspective began to clarify. Bashing the science is a good rhetorical strategy, but debating the science wasn't really the point. Opponents of GM crops are quick to accept scanty data without rigorous analysis because of their underlying antipathy to the corporate system through which these crops were developed. Genetically engineered crops themselves are not even the issue, but are a subset of more comprehensive and political opposition to large-scale farming and multinational corporations.

Once the obligatory science-trashing was done, each speaker went on to passionately express this more fundamental agenda, pitting sustainable agriculture against corporate agribusiness, small community-based businesses against multinational companies, indigenous cultures against the monolithic mass cultures of developed countries, and the tenets of technological progress against traditional values.

The pithy phrases ringing through the hall vigorously expressed an aversion to mainstream organizations and a deep distrust of institutional motives in promoting genetically modified crops: "We are seeing collusion between government and industry to foist genetically modified products on us." "Genetic engineering is the ultimate colonialism . . . Trade treaties are the gunboats employed by the new colonialism called globalization." "Genetic engineering is controlled by six gigantic corporations whose purpose is to control the global food supply." "We see a tremendous violation of collective rights of indigenous peoples."

The speakers also posed ethical objections to GM crops, stemming from a value system that emphasizes a quasi-religious respect for nature and a belief in traditional medical, agricultural, and cultural practices. Unlike the science and the politics, ethical issues are not easily argued because there are no facts that can be brought to bear that might modify opinion: "All this rests on a profound disrespect for creation." "Genetic engineering comes from a culture that views nature as an enemy that we must conquer, dominate, and subdue." "Its magic bullets offer no solutions to our real problems on earth."

Advances driven by science and technology had come too fast for the "Big Money, Bad $cience" participants. They feared the unknown side effects and consequences of biotechnology for human and environmental health, and did not believe in the benefits touted by an industry they did not trust. The people assembled in the Vogue Theatre were united by a belief that in moving ahead with GM crops we are losing some essential part of our human nature and planetary integrity, sacrificing close-to-the-earth lifestyles and values to the deities of technology, industry, and progress. Brewster Kneen has made a career out of voicing this yearning for bygone values, and he expressed his objections to GM crops simply: "One of the problems with biotechnology is that it only moves forward."

■■■ The North American opposition to genetically modified crops may still be an alternative movement, on the outside of the power structure looking in, but it is no longer at the fringe. It is made up of innumerable groups with tens of thousands of members, and has become as sophisticated as industry in its tactics, successful in using creative public relations to raise doubts among consumers, and influential in forcing some food corporations to abandon the use of genetically engineered crops in their products.

The opposition is not homogeneous, but made up of diverse groups, both large and small. Some present themselves as science based, such as the Union of Concerned Scientists, the Council for Responsible Genetics, and the International Center for Technology Assessment. Others, like Greenpeace and Friends of the Earth, include genetic engineering as part of a broader environmental agenda. Yet another segment, composed of groups such as the Organic Consumers Association, the National Farmers' Union, and the Pesticide Action Network, opposes GM crops because of their impact on sustainable agriculture. Others are more obscure but equally concerned, such as the moms in Mothers for Natural Law and Mothers and Others for a Livable Planet.

The flagship activist group in North America is clearly Greenpeace, which has set the tone of in-your-face environmental protest since 1971. Greenpeace is an international organization with its headquarters in Amsterdam, offices in over twenty countries around the globe, 253,000 members in the United States, and 2.5 million members worldwide. It began in 1971 with a handful of protesters from British Columbia, Canada, who sailed an aging fishing boat to Alaska's Amchitka Island in an effort to prevent the United States from testing a nuclear bomb. Their organization's mission soon broadened to saving whales and embarrassing polluters, and today Greenpeace members view themselves as using nonviolent, creative confrontations to expose broad global environmental problems and force the implementation of solutions.

Greenpeace in the United States has been conducting its anti-GM crusade since the 1980s, using attention-grabbing tactics, legal action, and mass advertising to raise public awareness. Its campaign has certainly irritated industry, and occasionally it has succeeded in forcing corporations to abandon genetically engineered products. Greenpeace's approach often comes across as self-righteous, with an "enough, already" feel to it, but one thing about Greenpeace activists cannot be denied: they are having more fun than industry's public relations people.

Kellogg's is one of the U.S. cereal giants currently under siege by the Greenpeace spin doctors. The company has annoyed Greenpeace because it promised its European customers not to use ingredients from GM crops, but continues to use GM corn in products sold in North America. In response, Greenpeace created FrankenTony, a genetically mutated version of Kellogg's symbol Tony the Tiger. FrankenTony appears at factories, media events, and the Cereal City, Michigan, headquarters of the company sporting a Greenpeace-modified box of Kellogg's Genetically Modified Frosted Fakes, labeled with phrases like "Untested" and "Hey, Kids, Get Your Gene Splicer." FrankenTony was joined at Christmastime by The Grinch Who Stole Breakfast, who visited Kellogg's main office with petitions and a shopping cart full of genetically modified corn. Greenpeace activists also flooded the e-mail of Kellogg's chief executive officer, Carlos Gutierrez, with seven thousand messages by encouraging visitors to the Greenpeace Web site to click on an icon that would send him an anti-GM letter.

As of November 2001, Kellogg's had not given in, but it only took a few weeks in 1999 for Gerber Baby Foods to succumb. Greenpeace took the clever step of having Gerber's Mixed Cereal for Baby analyzed. Given the high proportion of corn and soybean plantings that were genetically modified that year, it was not surprising that evidence of insect-resistant Bt corn and herbicide-tolerant Roundup Ready soybeans was discovered. The campaign against Gerber immediately moved into high gear with a blunt press release: "The Gerber baby isn't smiling today. Unlabeled genetically altered products leave parents little choice but to have their children used as guinea pigs in this corporate experiment with our food."

Gerber caved in quickly, promising not only to stop using genetically engineered corn and soybeans but also to use organic ingredients in its products whenever possible. Ironically, Gerber is owned by Novartis, which was one of the world's leading producers of genetically modified crops until it spun off its agribusiness section to form Syngenta.

Greenpeace and the other anti-GM activist groups also have become masters at creating advertisements and press releases that present seamlessly woven mixtures of fact and opinion. In 1999, for example, a coalition of more than sixty nonprofit organizations describing themselves as favoring democratic and ecologically sound alternatives to current practices took out a series of advertisements in the *New York Times*. Large type proclaimed: "Unlabeled, untested . . . and you're eating it . . . In secret, genetically engineered foods are showing up on American grocery shelves." Oddly, the ad also included a list of processed foods containing genetically engineered ingredients. If these foods were showing up secretly, the secret wasn't well kept.

The advertisement went on to accuse the U.S. government of refusing to require any safety testing, and the industry of producing products that violate the religious and ethical principles of millions of people. The text is replete with qualifying terms like "may" ("the genetic engineering of food may bring some undesirable effects") and "could" ("scientists warn that genetically engineered foods could produce a new allergen"), leaving the impression that what are generally considered to be remote risks such as toxic and allergic reactions, cancer, and suppressed immune systems are probable outcomes of genetic engineering.

I spent a considerable amount of time reading anti-GM advertisements, attending meetings like "Big Money, Bad $cience," visiting the innumerable Web sites that attack the GM food industry, looking at press releases, and talking with activists. I soon became numb, deluged by the cleverly crafted arguments of anti-GM spokespeople who wanted to sway my opinion rather than to provide an objective viewpoint. However, this feeling of being manipulated was no different from the feeling I experienced when talking to industry's spokespeople, although the opposition's public statements were more creatively composed and sprang from passion rather than the profit motive.

I spoke with Jenny Hillard, vice president in charge of issues and policy for the Consumers' Association of Canada, about the confusion engendered through what has become debate by sales pitch and slogan. We met at a coffee shop in a suburban shopping mall outside of Vancouver, British Columbia, where she was leading a focus group to obtain consumer feedback about biotechnology issues. In her sixties, Jenny is dynamic and passionate about promoting consumer rights, and she is constantly traveling within Canada and internationally to promote policies that favor consumers.

Her take on the quality of information available about genetically modified foods was not flattering: "I don't care whether it's Greenpeace or Monsanto or the government, we're not getting fair, accurate, reasonable information. We're either getting the Doomsday scenario or don't worry be happy, or we get the government saying we can't tell you anything, it's proprietary information. Lack of information, inaccurate information, biased information, one-sided information, no information . . . We think that's a big problem."

Industry and activist propaganda alike overwhelm the public with selected details and slanted interpretations that make it difficult to separate the facts about genetically modified crops from the ideology. The public relations strategy of industry has been to link reassuring lab-coated scientists emphasizing progress with traditional family farm values. The opposition has consistently nipped at the heels of industry, harping on the fear of unknown consequences, while also claiming the moral high ground of the same family farm.

In the end, it becomes a question of whose lobby is more effective. So far, critics are winning the public's attention while industry has triumphed in the regulatory arena.

Some of the opposition's strongest rhetoric about genetically engineered crops has been directed at the final food product. No adverse consequences have yet been found from eating food originating from any commercial GM crop, but the clever nickname "Frankenfoods" has nevertheless taken hold, becoming a rallying cry for protest against what critics view as a terribly dangerous disturbance of our food supply.

The propaganda spouted by both sides in this debate repeatedly invokes science to support or criticize safety testing. Industry and the regulatory agencies maintain that most food originating from genetically engineered crops is substantially equivalent to conventional food and that the testing conducted to detect toxic or allergenic substances is adequate. They claim that the science is clear and accessible, and that in the absence of any proven negative impact there is no justification for banning or limiting food produced from bioengineered crops.

The opposition invokes the classic arguments of the risk averse, that "substantially" does not mean "exactly" and that testing for allergens and toxins checks known but not unknown elements. GM opponents argue that lapses in food safety are plausible with GM crops, and that the potential for harm rather than actual case histories should be sufficient to at least delay crop registrations no matter how low the risk.

Food safety normally falls under the purview of the FDA, which tests food products if they have not previously or generally been recognized as safe on the basis of long-standing use. However, the EPA has regulated food produced from genetically modified Bt crops because these crops produce compounds that act as pesticides, and pesticides are under EPA jurisdiction.

Current food analysis protocols focus on determining whether novel substances would be harmful to the most sensitive consumers, infants and children. They test the product rather than the method used to produce it. Whether a crop has been genetically engineered or produced through traditional breeding methods is irrelevant; it is the characteristics of the food itself that are considered.

Even critics agree that the necessary methodology is available to determine whether GM food could potentially induce a toxic or allergic reaction. Tests include comparisons of the molecular structure of any proteins in the food with all known toxins and digestibility studies to confirm that the food will be broken down rapidly through digestion by stomach or intestinal fluid. Other tests involve feeding rodents various doses of the food to determine toxicity. In addition, scientists examine the proteins' mode of action to find out if they could bind in some harmful way to the human digestive system, and they also assess the potential to induce an allergic reaction by challenging immune systems with novel substances from new food products.

This system of testing is thorough and effective, and it is difficult to argue on any rational scientific grounds that a GM crop passing the full battery of these tests should concern consumers.

But it is not the testing protocol that worries opponents. Rather, they focus on the slippery term substantial equivalence and on regulatory decisions that have permitted restricted licensing when the product has not passed all safety tests.

Substantial equivalence means that a new product is similar to products already on the market and has no constituents that are not present in already approved foods, and therefore should not require further testing. Applying these criteria does assure the absence of known toxins or allergens, but critics of this policy insist that there could be subtle differences between already accepted and proposed foods, for example higher quantities of particular compounds or previously unknown mixtures of substances that could prove harmful when combined together in food. They disagree with the FDA policy that has allowed industry considerable latitude in arguing that a GM crop is substantially equivalent to its conventional counterpart and thus does not need to go through rigorous safety testing.

The weaknesses of the substantial equivalence policy were dissected in a January 2001 report commissioned by the Consumer Federation of America, a consortium of 270 pro-consumer groups. The report, *Breeding Distrust,* was compiled by Thomas McGarity and Patricia Hansen from the University of Texas School of Law, and came down hard on the science behind substantial equivalence: "Substantial equivalence is not a scientific theory; it is a debatable public policy that is currently driving risk management decisions in the United States. It is extremely vague and as such presents a weak foundation for a regulatory regime. Quite apart from its vagueness, expert-assessed similarities between a GM food and its natural counterpart provide little assurance that the GM food is safe for human consumption . . . The present testing regime in which a manufacturer or importer can avoid virtually all testing by simply finding that its product is substantially equivalent to existing food is not adequate to protect human health, and it is certainly not sufficient to secure public trust in the regulatory process."

FDA policy makers pondered substantial equivalence in trying to decide whether transgenic products should be classified as novel food additives. To date, the FDA has regarded most bio-engineered products as being comparable to substances found in conventional food and has not classified them as additives. Thus the FDA has been quicker to invoke substantial equivalence than the EPA, since crops secreting pesticides clearly are not equivalent to conventional crops. But although the FDA has been friendlier than the EPA to substantial equivalence, in practice the FDA sometimes does ask manufacturers to include reports on toxicity, changes in nutrients, and allergenicity for foods originating from GM crops. However, these consultations were voluntary until made mandatory in 2001, and the extent of the FDA scrutiny has been difficult to assess. The fact that the FDA places greater reliance on substantial equivalence than does the EPA worries critics, especially because most of the new crops under development today will fall under the more relaxed jurisdiction of the FDA.

Opponents of GM food also criticize regulatory agencies for granting a crop that does not pass scientific muster a temporary registration for a limited use until the manufacturer can submit more data that might clear the regulatory hurdles. This process is flagrantly pro-industry, and backfired on the EPA in the fall of 2000 when StarLink corn, a Bt variety approved for use only as animal feed, was found in taco chips destined for human tables.

StarLink was a Bt corn variety produced by Aventis when the company was still Aventis CropScience, and StarLink went through the regulatory system along with other Bt corns in the mid-1990s. The other companies' Bt products passed every test and were given five-year registrations, and StarLink almost made it through as well. But the Cry9C insect-killing protein used in the Aventis corn was unique to StarLink and failed one test for potential allergens, surviving for thirty minutes when exposed to digestive fluids. Although this finding does not mean that Cry9C will sensitize exposed individuals or cause an allergic reaction, its reduced digestibility suggested that it could. For that reason, the EPA licensed StarLink corn only for animal feed, and prohibited it from being used in human food products.

Millions of bushels of StarLink corn were planted by farmers, much of it under license to Garst Seeds, and in the fall of 2000 the Cry9C protein was discovered in taco shells and then in other corn products. In retrospect this is not surprising, because farmers had not been thoroughly warned by seed distributors that they needed to segregate StarLink corn from human-destined corns. Even for those aware of the EPA's limited registration it would have been difficult to restrict StarLink to animal feed under the hectic harvest and shipping conditions characteristic of most farms.

Aventis offered to buy back and destroy all of the StarLink harvest that year, at a cost of $100 million, voluntarily withdrew it from future production, and worked with farmers, grain elevator operators, flour mills, and food producers to identify commingled corn. The downstream economic damage was considerable. Kellogg's, for example, shut down its Memphis breakfast cereal factory for about a week to make sure that its corn supply did not contain StarLink, and Japan recalled much of the corn it had imported from the United States until the situation could be sorted out.

Aventis quickly realized that the task of tracking and identifying the StarLink corn that was moving through U.S. and foreign markets was daunting, and applied to the EPA for permission to let whatever StarLink corn was in the human food chain proceed into products. Their application included more data about StarLink's safety, and argued that there was no legitimate scientific reason to be concerned.

This report and other data were considered by a scientific advisory panel convened by the EPA on 28 November 2000. The panel members concluded that the potential for allergenicity from StarLink in the food chain was low, because of the minute level of StarLink corn found in products and the lack of clear data demonstrating that it could cause an allergic reaction. They did not, however, let Aventis off the hook entirely, blaming the company for not conducting tests that would directly test for allergic responses: "The Panel members were uncomfortable with the available data; there was an expectation of more . . . These points were made in earlier Scientific Advisory Panel meetings, and have not been addressed."

They also urged government regulators to follow up on reports concerning twenty-eight individuals from across the United States who might have reacted to corn products that autumn. These patients had experienced mild to severe allergic reactions, and corn products were among the suspected foods eaten prior to onset of their reactions. The panel strongly suggested that the corn be tested for StarLink's Cry9C protein, and that blood serum from the reactive individuals be examined for sensitivity to Cry9C. These studies were conducted respectively by the FDA and the Centers for Disease Control and Prevention (CDC), and their reports were issued in June 2001. The FDA study found no evidence for the presence of Cry9C proteins in any food consumed by the patients, and the CDC report similarly concluded that none of the blood serum samples from those individuals reacted in a manner consistent with their having experienced an allergic reaction to Cry9C protein.

The StarLink situation and arguments about substantial equivalence beg the question "Are food products made from GM crops safe?" The scientific answer promoted by industry and accepted by regulators so far has been a clear "yes." There have been no proven health-related problems from GM foods to date, and the StarLink incident was caused by regulatory laxity rather than problems with analytical methods. In hindsight, the EPA should have recognized the potential for Starlink corn to move from animal to human markets and followed its own scientific guidelines by not registering StarLink for any use until allergenicity had been ruled out.

The point-counterpoint arguments by critics and proponents on issues such as substantial equivalence reflect differences in scientific opinion concerning how much analysis is sufficient. Given the public fever spiked by terms like "Frankenfoods" and aggressive campaigns by groups such as Greenpeace, it would seem prudent to take the safest path and test all GM foods directly rather than relying on their equivalence to conventional products. The time and expense required to do so means that products may take a bit longer to get to market and development costs may rise slightly, but that seems a small price to pay for consumer security and customer confidence.

The food-testing methods currently available to regulators are not "Bad $cience," and it is important to recognize that there has yet to be a single credibly supported incident of GM foods adversely affecting human health. Nevertheless, the public perception is that the FDA and EPA are beholden to "Big Money" and are not applying strict enough testing criteria. The available scientific methods can provide reasonable security about the safety of genetically modified foods, and perhaps regulators need to reassure the public by conducting even more stringent testing and registering only crops that pass with flying colors. Policies such as the acceptance of substantial equivalence and incidents such as the discovery of StarLink in human food have not bolstered the already shaky public confidence in genetically modified crops.

The strident voice of agricultural biotechnology critics may have left a legacy of confusion in the wake of its anti-GM campaign, but it has stimulated mainstream consumer advocacy organizations to campaign for full labels on GM food. Nonprofit consumer groups have been lobbying government about labeling food products containing bioengineered foods, but industry has been successful at persuading regulators to resist this agitated and insistent demand to give people the simple right to know what is in the food they buy.

The first rule of thumb in consumer advocacy is transparency, that consumers are entitled to full disclosure about product qualities and risks and that government and industry have the responsibility to provide the relevant information. Consumer associations are as baffled as the general public about hazards in food from genetically engineered crops, and their solution to the confusing barrage of mixed messages from critics and advocates has been to demand labeling so that customers can make their own choices.

Carol Foreman, director of the Food Policy Institute of the influential Washington-based Consumers Federation of America, highlighted the public interest in an April 2000 speech before a colloquium on science and technology put on by the American Association for the Advancement of Science: "Given the important role of food in our lives, it isn't surprising that the population tends to be extremely risk averse and not always rational about food. Now we may eat too much or choose to gamble on a steak tartare, but most of us are unwilling to tolerate any food safety risk that is imposed on us by someone else . . . Genetically engineered foods are invisible in the retail product and not labeled, so consumers don't have the option of choosing to avoid them."

The federation's 2001 report *Breeding Distrust* transformed Foreman's concern into a consumer's call to arms: "There appears to be an emerging worldwide consensus that GM foods should be labeled to allow consumers to choose whether to purchase them . . . The reasons for requiring labeling of GM foods are so compelling that the matter should no longer be voluntary. Public acceptance of GM foods requires that consumers be given a choice."

Despite intense lobbying and numerous public opinion polls indicating 80 to 90 percent support from the North American public for mandatory labeling, the FDA has refused to accept the pro-labeling arguments. Much of its reluctance stems from the fact that in the convoluted world of regulators accepting labeling would be to admit that bioengineered food might indeed be different from conventional products, thereby invalidating its substantial equivalence policies and opening the door to more extensive food testing.

At the same time the Consumers Federation issued its report demanding food labeling, the FDA issued a report shutting the door on mandatory identification of GM products: "The FDA has no basis for concluding that bioengineered foods differ from other foods in any meaningful or uniform way, or that foods developed by the new techniques present any different or greater safety concern than foods developed by traditional plant breeding . . . The comments that addressed labeling requested mandatory disclosure of the fact that the food or its ingredients were bioengineered. These comments did not provide data or other information regarding consequences to consumers from eating the foods or any other basis for FDA to find that such a disclosure was a material fact. The comments were mainly expressions of concern about the unknown. The FDA is therefore reaffirming its decision to not require special labeling of all bioengineered foods."

Even if implemented voluntarily, food labeling can be fraught with problems, ranging from whether a label should communicate presence or affirm absence to questions about exactly what a label about genetically modified foods should say. The FDA report denying mandatory labeling for GM products attempted to placate label advocates by providing guidelines for voluntary labeling, but its approach focused on issues of label accuracy rather than on the desirability of proceeding with labeling.

For example, the report presents a rambling, lengthy, and bureaucratically convoluted discussion about using the negative label "GMO free"—free of genetically modified organisms. The word "genetically" is criticized because all crops have been genetically selected, and a bioengineered component is just one of many types of genetic input. The term "modified" is equally vague according to the FDA language police, since modification could refer to any alteration in food composition. Finally, most foods do not contain organisms, so to say a product is free of organisms has no meaning.

Once again the science has fallen victim to language, this time wrought by biotechnology regulators rather than critics. Clear labeling based on solid science has been developed for innumerable food traits, including nutritional content, potential allergens, such as peanuts, and recommended daily intake. It would take some effort to develop unambiguous labels for genetically engineered food, but the problem is far from insoluble. Rather, it has been the unwillingness of government and industry to bend in the face of consumer unease that has been the most insurmountable problem. Their reluctance to fully test and clearly label genetically modified organisms has reinforced the opposition's worst fear, that multinational corporations are colluding with government regulators to hide important information about biotechnology.

The irony is that the science of food analysis when fully practiced has yet to reveal any major faults with GM food, and the ability of thorough testing to prevent problem foods from reaching the market should be industry's best safeguard against bad publicity. The public would be more receptive to bioengineered foods if industry proudly labeled the benefits of its attributes rather than attempted to hide the fact of its existence.

The opposition campaign against genetically engineered crops has been remarkably successful given the absence of clear, substantial, and demonstrated negative impacts on environmental or human health. The critics have used unjustified interpretations of a few ambiguous and preliminary scientific findings, clever public relations gimmicks, and the ubiquitous fear of unknown consequences from new technologies to mount an effective assault on biotechnology. Today, many North American food manufacturing companies have eliminated or are considering eliminating GM crops as sources for their products, and poll after poll has indicated public confusion about the safety of GM foods and lack of confidence in government regulators and bioengineering firms.

Of course, critics are correct in demanding that proper science be conducted before a company can proceed with a new bioengineered product. Eventually we may discover that a novel GM crop can indeed drive monarch butterflies to extinction, cause life-threatening allergic reactions in humans, or pass genes from bacteria to plants to humans. But although all these things are possible, the preponderance of the data does not support these and other horrific consequences for the bioengineered crops that are currently available.

Given this so-far clean bill of health for bioengineered crops and their positive economic benefits, the campaign against genetically modified organisms should not have been so successful. A mutated Tony the Tiger is a clever public relations ploy, listing GM-containing foods commonly eaten by infants and children is an excellent tactic, and cute titles like "Big Money, Bad $cience" and "Frankenfoods" are media magnets. Yet the proponents of biotechnology have the vast public relations resources of industry and government on their side to counter the critics.

It is more the blatantly self-serving attitudes of industry that have eroded public confidence in genetically modified crops, creating an atmosphere where the tactics of the opposition have flourished even in the absence of an actual environmental or health disaster. Industries producing GM products have lobbied hard to get the government to accept the short cut of substantial equivalence, thereby saving themselves time and money in bringing products to market but opening the door for critics to argue that they have cut corners in food safety testing. Businesses have been successful at persuading regulators to accept their own studies rather than hiring independent laboratories to provide data, causing a distrustful public to discount the reliability of industry research. Manufacturers have succeeded in preventing mandatory labeling but in the process have lost the trust of consumers who want the right to choose their own risks.

The critics may be shrill, but their opposition is rooted in the most powerful force in public debate, the issue of trust. The failure of the biotechnology industry to win the public's confidence has increased the likelihood of protest and created an atmosphere conducive to dissent. Corporations have responded by repeating their reassurances that the science is fine, maintaining that their proprietary data shown in secret to regulators should provide sufficient proof of the safety of GM crops, but this response clearly has not been sufficient.

Belatedly, industry is trying to catch up to its critics. In a remarkable mea culpa speech delivered to the Farm Journal Conference on 27 November 2000 by Monsanto's chief executive officer Hendrik A. Verfaillie, he confessed his company's sins and promised to do better: "My company had focused so much attention on getting the technology right for our customer the grower that we didn't fully take into account the issues and concerns it raised for other people. We missed the fact that this technology raises major issues, issues of ethics, of choice, of trust, even of democracy and globalization. We didn't understand that when it comes to a serious public concern, the more you stand to make a profit in the marketplace, the less credibility you have in the marketplace of ideas. When we tried to explain the benefits, the science, and the safety, we did not understand that our tone, our very approach, was seen as arrogant. We were still in the 'trust me' mode when the expectation was 'show me.' Instead of happily ever after, this new technology became the focal point of public conflict."

Biotechnology opponents have grasped one simple public relations fact considerably better than industry. Science and data cannot substitute for actions and statements that engender trust. Unless industry wins back public confidence, it will be FrankenTony rather than fields of pest-resistant corn that we remember as the legacy of genetic engineering.

> Canola lends itself to biotechnological tinkering
> more than any other plant.
>
> **National Research Council of Canada,** *From Rapeseed to*
> *Canola: The Billion Dollar Success Story,* 1992

Saving the Family Farm

Percy Schmeiser plays well to the camera, and has been much in demand lately as the stereotypical angry farmer opposed to the multinational seed companies. At one point he was doing three hundred interviews a month for media around the world, and television crews from as far away as Japan and Denmark still arrive regularly to film documentaries about Schmeiser's David and Goliath battle against the Monsanto Company.

Schmeiser is a dapper farmer in his seventies from the small town of Bruno, Saskatchewan, about sixty miles east of Saskatoon. He is well suited to his role as giant killer. Percy talks readily and in an effortless style developed during his many years of representing his town as mayor and provincial legislator, but he also has his feet firmly in the furrow.

Schmeiser farms, as did his immigrant grandfather and father before him, and like many farmers he is struggling today to keep the family farm alive. In the last few years, however, most of his time has been spent off the farm talking to the media or his lawyer, because Percy Schmeiser and Monsanto sued each other for millions of dollars. Their dispute exemplifies the issues that genetically modified crops pose for contemporary farmers and seed companies.

The core of the litigation was Monsanto's claim that Schmeiser used seeds of its genetically modified Roundup Ready canola variety illegally, and Schmeiser's counter-suit that Monsanto defamed his character. Schmeiser's version of events was compelling. He said that Monsanto employees sneaked onto his land, sampled his crops, and found a few plants of their herbicide-resistant canola that had sprouted from seeds inadvertently blown onto his land, or that had developed because wind-blown pollen had fertilized his conventional canola varieties with the gene resistant to Roundup. Monsanto put forward another version of what happened. Its lawsuit against Schmeiser claimed that he harvested seed from these fields, and in replanting it the following year violated Monsanto's patent.

It's a classic story, with issues of patent rights clashing with the individual farmers' attempts to prevent the demise of the family farm. However, most canola farmers are not as agitated as Schmeiser is about GM crops, and his lawsuit was generally viewed as a side issue detracting from the main question in farming today: How can a farmer make a living growing crops for an intensely competitive and unfairly regulated world market, and what agronomic tools are available to reduce costs while increasing yields and profits?

The Canadian prairies are a good place, and canola an interesting crop, for exploring the contemporary issues raised by farming and genetic engineering, because canola was created when Canadian scientists manipulated the genome of its predecessor, rape, through traditional plant breeding methods. Canola exists only because of biotechnology, although it is technology from a time earlier than our current recombinant DNA era. Canola is a great Canadian success story, a combination of brilliant agricultural research and an economic climate in which prairie farmers needed to diversify to preserve their farms. Today, canola farmers are in trouble, and they once again are looking to biotechnology to save the family farm.

Canola did not exist at all before the 1970s. Some rape—a crop that is closely related to cabbage, cauliflower, Brussels sprouts, broccoli, and turnips—had been grown in Canada since the early 1940s. Commercial cultivation of the oil-rich rape seed had begun during World War II, when plant oils were desperately needed for food and industrial applications, especially as types of cooking oil and as components of fuels, lubricants, soaps, and synthetic rubber. However, the oil from rape seeds never caught on as food. For one thing, it was dark, dank smelling, and foul tasting, but rape oil also presented another problem. In tests with rats it induced weight loss, enlarged adrenal glands, caused sterility, and piled up fat deposits around the hearts and kidneys, and thus potentially could be harmful for humans.

Canola was developed primarily by government scientists from the National Research Council and Agriculture Canada during the 1950s, 1960s, and 1970s, when working for the government was still considered prestigious. Their work was stimulated by a congruence of political and economic events. Prairie farmers were taking a beating on world markets because of a glut of their major crop, wheat, and they desperately needed to diversify into some alternative crops. At the same time physicians and the public were becoming aware of the benefits provided by a diet high in unsaturated rather than saturated fats. Government investment in research was increasing, providing attractive opportunities for young, ambitious scientists.

The technical issues involved in the invention of canola were daunting. Photographs from that period show exclusively male scientists dressed stiffly in suits, white shirts, and ties, standing in front of towering analytical machinery with surfaces crammed full of knobs, dials, and gauges, retrospectively looking like caricatures of scientists and laboratory equipment from a bad 1950s science fiction movie.

Their first challenge with this then state-of-the-art equipment was to develop techniques to analyze the oil and protein components from rape seeds. This was not a simple task, since rape oil is a complex mixture of elongated oil and protein molecules. It was a bit like taking a sample from minestrone soup, grinding it to a puree, pouring it into a black box, and producing a readout at the other end telling you what vegetables were in the pot. These analyses were almost impossible to perform, but serendipitously the invention of the gas-liquid chromatograph (GLC) enabled the scientists working on this project to make some progress. This analytical device carries volatile components in helium gas along a thin tube, and since each component moves at a slightly different rate, a sensor can detect distinct peaks of specific compounds, which then can be identified.

Analysis of rape oil was one of the first processes to benefit from the GLC, and it revealed that the oils contained between 25 and 40 percent erucic acid, which turned out to be the molecule responsible for inducing health problems in rats. The next step was to breed plants that had lower levels of erucic acids by selecting and breeding varieties with genes coding for high levels of molecule-shortening enzymes that would transform the long erucic acid molecules into shorter and more beneficial oleic acids.

At this point the plant breeders took over from the chemists, and developed a brilliant and novel breeding method that allowed them to make quick progress. They worked out a technique by which a seed could be cut in half, one half of the seed could be analyzed by the chemists, and the other half of the seed could be grown for breeding purposes if the tested half had a low level of erucic acids. Over many years they were able to select the erucic acid content down to less than 5 percent, then 2 percent, and eventually virtually 0 percent.

Another problem was that the residue remaining after the oil was crushed from the seeds was high in substances called glucosinolates, compounds that induce thyroid disorders in livestock. The use of the by-product meal for animal feed was an important issue, because sales of oil seed would not have provided sufficient income to justify the new crop unless the buyers could produce and sell the meal. Again, careful selection resulted in the reduction of glucosinolates to acceptable levels in the meal produced as livestock feed. Then, a classic but dazzling series of crosses between low-erucic acid and low-glucosinolate lines created double-low varieties containing almost none of either substance.

Every step of this process broke new scientific and plant breeding ground, and in 1974 the first plants of the new crop were approved and licensed for commercial agriculture. This offspring of rape needed a name, and "canola" was chosen, both as a composite of "Canadian" and "oil" and because it rhymed with that healthful breakfast cereal granola. Today, canola's bright yellow flowers have become a trademark image of the Canadian summer prairie, and canola has become one of Canada's most important crops, rotating with grains and legumes in farmers' fields. Current production is valued at $2 billion (U.S.) annually, with 7.8 million tons of seed produced on 13.7 million acres of prime Canadian prairie farmland. Canola oil is prized by consumers because of its low level of saturated fats—the lowest of all the edible oils, about one half the levels in corn, olive, and soybean oils. In addition, it is high in linoleic acids, and especially oleic acids, both of which are considered healthy constituents of the human diet.

There is irony in canola's success, because there is a glut of canola seed on the world market today, and low prices once again are forcing farmers to turn to biotechnology to provide diverse options for crop production and marketing. And the canola market is depressed at the very time when family farming on the Canadian prairies is again threatened with extinction.

All farming is in trouble in North America, because of increasing costs, rapidly fluctuating world markets, heavily subsidized government support for many commodities in other countries, and trade agreements that increase commerce but can cripple the sale of particular crops as policies benefiting one commodity are traded off against those favoring another. These problems are all outside of a farmer's control, and in spite of considerable lobbying they remain in the realm of politics and government rather than the realm of farm management. What farmers can control is the crop they plant, and most farmers believe that the best way to preserve their farms, values, and traditional lifestyle is to embrace yet another new technology and grow genetically modified crops.

Canola has proven to be an excellent crop in which to practice the developing craft of genetic engineering. It readily accepts genes from bacteria, other plants, and even animals, and is therefore an ideal medium for introducing genes carrying useful traits. Herbicide tolerance was one of the first traits transferred into canola in the 1980s, because of the economic significance of weed control and the availability of relevant genes from bacteria. In the year 2000 over 80 percent of growers planted one of the genetically modified, herbicide-tolerant canola varieties on 55 percent of Canada's canola acreage.

The concept of herbicide tolerance is simple. Most herbicides deactivate proteins that are found only in plants, but these proteins generally are found in all broad-leaved plants, and thus most herbicides will harm crops as much as weeds. If a gene that confers resistance to herbicides can be found and transferred into a crop, then a weed-killing herbicide can be sprayed while the crop is in the field. Farmers can determine how serious their weed problem is after the crop is planted, spraying only when necessary and avoiding prophylactic sprays before planting or after the harvest. Thus the use of GM canola can provide better weed control while reducing herbicide use and the cost of crop production.

Glyphosate, the active ingredient of Monsanto's Roundup herbicide, interferes with amino acid synthesis in plants by blocking the action of an enzyme. Monsanto's scientists were able to induce mutations in genes from a common bacteria found in soil. These mutated genes produce higher levels of the enzyme that the herbicide targets or else contain a modified form of the enzyme that the herbicide does not affect. Scientists cut these genes out of the bacteria and pasted them into canola, creating the herbicide-tolerant Roundup Ready canola that became available to farmers in the mid-1990s and that has become a common variety used by growers on the prairies.

████ It is not surprising that Canadian farmers were among the first to adopt this innovative technology. The prairie provinces of Alberta, Saskatchewan, and Manitoba make up the principal canola-growing region in Canada, and even the existence of farming on the Canadian prairies is a minor miracle. The climate is truly harsh, at the northern limits of agriculture, where temperatures of −35°C and bitterly cold winds are common throughout the long, dark winters. The summers also can be formidable, with high winds that scatter improperly tilled soil across the provinces, unpredictable but sparse rainfall, and biting flies that make a trip outside intolerable any time near sunrise or sunset.

Prairie agriculture exists in this region for many reasons: Up to twenty hours of daylight during the summer months makes for excellent crop growth; there are fewer insect pests than are found further south; and, most significant, the immigrant farmers who colonized the prairies and their contemporary descendants were and are extraordinarily resourceful individuals. I have spent a considerable amount of time with Canadian farmers over the years, and I would characterize the typical prairie farmer as intelligent, articulate, measured, modest, patient, and unusually generous, with a strong sense of family and heritage.

Farmers have another common trait, concern about the future of their farms and of their families. Their degree of anxiety ranges from uneasiness to full-blown angst, but it is almost impossible to have any discussion on the prairies today that doesn't eventually touch on predictions of the death of the family farm. Most farms today are still family run, but the writing is on the wall. Across Canada, the number of farms dropped about 2 percent a year during the 1990s, and average farm acreage increased by about the same amount annually. In 1996, the last Canadian census year, one third of farms produced over 80 percent of Canada's crops, and only 4 percent of farms produced 38 percent. Further, about half of the small farms lost money, while only 10 percent of the large farms were unprofitable. Figures from the United States are comparable.

The trend is clearly toward fewer and larger farms, with the size of each increasing to a magnitude beyond the ability of one family to operate. The economies that only a large scale can bring about are driving farming into a corporate rather than a family mode and toward a future in which a few large corporations own big farms worked by salaried employees. Family farmers are trying to reverse this trend for the next generation, but they are fighting an uphill battle, attempting to combat the developing reality that the economics of farming on the Canadian prairies soon may not allow a family to live and work on their own farm. Some agricultural experts believe corporate farming will result in improved agricultural production or profits, but few on the prairies welcome this drastic change in their lifestyle.

Almost any farm-related issue on the prairies today eventually is linked to how it affects the survival of family farms, and genetically modified crops are no exception. Prairie perspectives on GM crops have diverged into two opposing views. First is a pragmatic belief that GM crops are a useful and important tool that can improve yields and reduce costs; as such, they can play an important role in making smaller farms more profitable and thus saving the family farm. The other view, more controversial and much more publicized, is that biotechnology concentrates the means of production in the hands of a few multinational conglomerates, and will accelerate the death of family farming.

Percy Schmeiser has been singled out by the media as the spokesperson for farmers on issues about seed ownership, patent rights, and the influence of multinational conglomerate companies. These questions are important, but they are different from those of urbanites who have their own fears about the safety of genetically modified crops. Schmeiser's agenda, and that of farm organizations like the National Farmers Union, have much to do with farmers' independence and the fate of the family farm, and very little to do with the issues of environmental protection and food quality that concern consumers worried about genetic modification.

I spent an entertaining afternoon and early evening talking with Schmeiser, or rather listening, since being with him is more like watching a one-man play than participating in an interactive discussion. Schmeiser is intensely animated and deeply committed to his crusade for farmers' rights. His enemy is any large company that he thinks puts its own interests above those of farmers.

The core of Schmeiser's belief system is his conviction that farmers should be able to produce their own seed and reuse it the next year. This may seem like an odd basis for a crusade, but it is a principle to which he adheres with passion: "Our whole culture has been built up that you grew a plant, you produced seed, you used some of that seed for the crop next year. That has been going on for countless centuries, and I see this as a real threat to the whole farming industry and the individualism and the rights of individual farms to do what they want."

Seed ownership is governed by patent law, and the bottom line in Canada and the United States is that a seed company that creates a genetically modified variety can patent it, and can expect protection under patent law against any infringement of that patent. This frightens Schmeiser: "Because they put this one gene into a seed, they're claiming ownership of the whole seed, even though the seed contains thousands of other genes developed by man and nature . . . so here's the big scare now: if Monsanto can put a gene into any plant life, whether it's trees, plants, animals— and what about human beings?—it's their property, and that's why this case of mine has become such a landmark case to decide how far anyone can go with the patenting of a life-giving form . . . That's pretty scary, that's getting back to farmers' rights."

Monsanto has also drawn fire from Schmeiser and his supporters because its technical use agreement (TUA) contains clauses that rub some farmers the wrong way. The TUA that growers must sign when they purchase Roundup Ready seed prevents them from saving and reusing the seed next year, gives Monsanto the right to come on their property and inspect the seed for up to three years, provides for a $10 per acre fee above and beyond the seed cost, and limits the farmer's right to speak publicly about any subsequent conflicts with Monsanto. The company, of course, insists that its seed is proprietary and cost them hundreds of millions of dollars to develop; and hence, Monsanto says, the TUA simply protects its interests.

The TUA and patent rights conflicts are exacerbated by the perception among Schmeiser and his supporters that Monsanto is an overbearing multinational company, and rumors abound about the sleazy side of its patent protection activities. Schmeiser has at various times alleged that Monsanto hired a detective firm to ferret out farmers who use Roundup Ready seed illegally; instituted an incentive program in which the company gave a leather jacket to any farmer who turned in his neighbor; and paid spray planes to fly over canola fields of farmers who had not purchased Roundup Ready seed, spot-spraying Roundup to determine if any resistant plants were in the fields. Monsanto won't confirm or deny these allegations, but whether these incidents happened or not, they reflect how contentious the interaction between Schmeiser and Monsanto has become.

Ultimately, the lawsuits between Schmeiser and Monsanto are rooted in the populist prairie tradition of farmers' independence, and in the lifestyle of the family farm. Schmeiser is most earnest when he goes back to his roots: "A lot of us farmers in my age group don't know who's going to take over our land. Our kids don't want to go through what they saw their parents go through. I look back at my grandfather and my father who really struggled to make a living here. They came over from the old country just to get away from this type of control. It's wrong what they're doing, it's just wrong, and that's why I'm fighting it."

Noble words, but the judge didn't see it that way. Canada's Federal Court in Saskatchewan ignored Schmeiser's emotional family farm issues and decided in favor of Monsanto in 2001. The court's ruling stated that it was irrelevant whether Roundup Ready canola originally was planted or blew onto Schmeiser's fields. Either way, he knowingly replanted and grew Monsanto's seeds in subsequent years without its permission, thereby violating Monsanto's patent.

The biotechnology industry was ecstatic at the court's vindication of GM crops as legitimate intellectual property, but Schmeiser was devastated. He is continuing his battle, but the court's decision was a blow: "I've lost fifty years of work because of a company's genetically altered seed getting into my canola, destroying what I've worked for, destroying my property, and getting sued on top of it."

Schmeiser sees himself as a lone voice crying out for justice, but there also are more organized voices speaking out for farmers' rights and the family farm in Canada. Most prominent is the National Farmers Union (NFU), which represents ten thousand family farms under the slogan "A fair shake for the farmer." NFU's head office is in a nondescript industrial area at the edge of Saskatoon, in a shabby metal-clad building that would not be out of place on any family farm. The media-savvy executive director, Darrin Qualman, wears crisply pressed three-piece suits that would fit naturally in any boardroom, but his look seems out of place in an office filled with worn desks and scarred, overflowing file cabinets.

His message is delivered in carefully crafted sound bites, yet is anything but corporate. The NFU perspective fits neatly into the social democratic tradition that has always been an important part of Canadian prairie politics. Nevertheless, even the left-leaning NFU is officially ambivalent about genetic engineering, because a good proportion of its members grow genetically modified crops. This is a group that takes a stand on every issue even remotely connected with agriculture, and it does have a position paper on intellectual property rights and genetic engineering.

The NFU begins listing its fundamental principles on this subject with the vaguely anti-GM statement that "the delicate balance of life is upset when humans manipulate parts of it," but the remainder of its policies have little to do with perceived or real dangers from genetic recombination and a lot to do with the fear that corporations will control the food supply. The NFU views the negative consumer response to GM food as useful, not because the NFU opposes genetic engineering but because consumer reactions might hurt corporate interests. Genetic modification is fine, with the caveat that "citizens and their governments, not corporations, must control genetic engineering. The products of genetic engineering, all forms of life and components of life, must remain in the public domain."

Still, Qualman has a hard time sticking to the official NFU policy, and his thinking, like Schmeiser's, is rooted in an overriding concern for the future of family farming. For Qualman, genetic engineering is not the way to go: "I personally am opposed to GM crops, but our membership isn't. All the evidence is that technology will not save farming. There is a good reason to believe that biotechnology will make farming less profitable. On the question of ban it or don't ban it, it's very hard to come to a final conclusion. Canada hasn't had the debate yet, but if Canada does decide to engage in that debate, in the end we will reject genetically modified food."

Schmeiser's crusade and the somewhat more ambiguous position of the National Farmers Union are seductive. Who doesn't like to criticize multinational corporations, and blame the conglomerates? Yet, as I spoke with more and more farmers a different picture began to emerge, one reflecting the fundamental dilemma GM crops pose for canola growers, and indeed the entire farming community. Unable to control subsidies, trade agreements, and input costs, farmers can control only how they grow their crops, and most farmers believe that the best way to preserve their farms, values, and lifestyles is to embrace technology rather than to retreat into an idealized version of yesterday's family farm.

I spent two winter weeks traveling across Saskatchewan and Manitoba after meeting with Schmeiser and then Qualman, visiting with canola farmers in cities and small towns to talk about farming and biotechnology. During the day, I drove across great expanses of land covered in dirty-white snow, with the sky above a dull gray or pallid blue during the few hours of daylight. But the night skies were clear, sprinkled with innumerable stars, and one night driving from Winnipeg to Brandon, Manitoba, I saw the northern lights shimmering in a dazzling display of beauty.

We met in generic coffee shops and on the farms, over endless cups of weak coffee. This is not Martha Stewart territory; ersatz country charm is not welcome here, and I was surprised by the pervasive eighties décor in the motels, restaurants, and homes. I never visited a homestead that fit my picture-book image of an old farmhouse, or stayed in a quaint country inn. Double-sided trailers were the norm on the farm, and I slept in motels with bland rooms overlooking indoor atriums containing tiny swimming pools and hot tubs, surrounded by a few tropical-looking artificial plants.

But the people themselves were very hospitable. In farming communities, everyone knows everyone else, even in these two large provinces, each the size of a small country. I spent an afternoon with friends in Saskatoon who are not farmers, but when I mentioned Percy Schmeiser they told me that they had rented their first house in Regina, Saskatchewan, from his son. When I said I had visited the National Farmers Union, they mentioned that Qualman's wife ran the bookstore two blocks over. Saskatoon bills itself as a community having "big city amenities with a small town spirit," and that's pretty much the attitude in every prairie town.

All of the prairie farmers I met with extolled the virtues of their own family farm heritage, but none of them shared Schmeiser's and Qualman's view that Monsanto's policies concerning GM canola either are wrong or threaten the sacred family farm. The growers I spoke with consider themselves professional producers, and while they admire Schmeiser's and the NFU's spunk and independent bent, they don't agree with them. They see Roundup Ready and other GM crops as a means to help bring in the crop economically, and as the type of innovative tool that offers the only real opportunity to preserve their lifestyle.

The first point at which most canola farmers diverge from the antibiotechnology movement is that they have little interest in the traditional farming practice of saving seed to reuse the following year. Few canola farmers still save seed, because it is more economical to buy seed from specialty companies that excel at selecting and producing certified seed. Buying selected seed has many advantages, not the least of which is the increased productivity of hybrid varieties. Like corn, hybrid canola is created by careful planting of male and female canola lines in adjoining rows, ensuring the cross-pollination that results in higher-yielding hybrid seed for farmers to plant in their fields. This is true for both conventional and GM varieties, and better production from cross-pollinated seed more than compensates for the extra cost.

Buying seed annually has other advantages. Certified seed comes cleaned, meaning that the weed seeds which inevitably get mixed in during harvesting have been removed. Also, certified seed is pretreated with coatings that reduce attacks by fungi and insects. Finally, the science of selecting canola with particular traits has become so sophisticated that no individual farmer can hope to develop new hybrids, and most farmers are more than happy to let specialists deal with creating and growing seed for more productive varieties.

Most farmers also see many advantages to the GM varieties. Producers view canola varietal development as a historical continuum, ranging from early and traditional selection for a trait found within the plant, to current techniques of transferring bacterial genes conferring herbicide resistance into the plant, to the future creation of even more productive types of GM crops. For farmers, the focus is on the outcome rather than the method, and they lump all plant-breeding methods together as selective tools rather than regarding GM plants as radically new and dangerous technologies.

The most important advantage of the GM canolas being sold today is improved weed control, accompanied by reduced herbicide use. Weeds are the single most important production problem facing canola farmers, because canola is not a strong competitor in its early growth stages. Prairie weeds include plants such as Canada thistle, cleavers, lamb's quarters, smartweed, wild mustard, and wild oats, in addition to volunteer seeds from cultivated grain crops that often blow into a canola field and cause problems the following season. Producers have fought against weeds by tilling, selecting varieties that compete well with weeds, rotating canola with other crops such as grains and legumes, and using a shifting array of herbicides.

The development of genetically modified canola that is herbicide resistant has allowed spraying against weeds while the crop is in the field, but the key issue for producers is whether GM crops result in higher profits than do conventional crops after the additional costs of GM varieties have been deducted. To examine this and many other production issues, the Canola Council of Canada established Canola Production Centres, farm locations across the prairie provinces where researchers could rigorously and independently field-test varieties against each other. Conventional crops have been compared to the GM crops in these tests, and the genetically modified canolas often out-perform the nonresistant varieties. Roundup Ready, for example, can double the profit of canola production, an obviously compelling result for producers operating on a very narrow margin.

Not only are genetically modified varieties economically beneficial, but they fit well with the prevailing tendency among farmers of adopting any cost-effective practice that can reduce chemical use on the farm. Contrary to city dwellers' stereotypes, farmers are averse to using pesticides, both because of their cost and because of a real concern for their own and their family's health, as well as for environmental safety and the reduction of residues in food. Most producers I spoke with who are using herbicide-tolerant canola thought that it was reducing their need to apply herbicides, but there have yet to be good quantitative studies to verify this effect for canola.

The use of GM crops has other advantages such as decreasing soil erosion, because they allow for direct seeding and reduce the need to till fields before planting or after harvesting in order to disrupt weed growth. Finally, herbicide-resistant GM canola today comes in many types, each resistant to a different herbicide, and by rotating these varieties growers can reduce the likelihood that weeds will become resistant to any one of the commonly used herbicides.

Canola producers also differ from Schmeiser and Qualman in their attitudes about the heavy-handed practices of multinational companies such as Monsanto. For one thing, virtually every farmer I spoke with pointed out to me that there are ten companies selling over two hundred varieties of canola seed in Canada, and any producer disgruntled with a particular corporation has the option of shopping elsewhere. Also, most varieties of canola have always been sold under license agreements, and so a seed company patenting a GM variety is not a new phenomenon for producers. Mega-conglomerates like Monsanto may be obvious targets for those with an ingrained distrust of industry, but vehement attacks on Monsanto by farmers are not common.

Even Monsanto's technical use agreement is not objectionable to mainstream canola producers. Jonothon Roskos, a young and politically active canola grower from Dufresne, Manitoba, told me, "It's a contract, it's no different than delivering grain to the elevator on a delivery contract. You're not being held ransom."

Bruce Dalgarno, an astute Newdale, Manitoba, farmer and a former president of the Canola Council, the Canadian industry organization that oversees research and marketing, has a similar perspective: "Roundup Ready sales have risen astronomically over the last four years. Some of the guys are certainly ticked off, but I would think it's primarily a few farmers that get bent out of shape . . . Farmers say, yeah, as long as I can make a dollar at it, it doesn't really matter who has got the control over it."

Farmers are not overly worried about the environmental impact of GM canola, either, because they view environmental issues as management rather than ethical problems, and believe that any deleterious impact from GM crops can easily be managed. The principal environmental problem with a crop like Roundup Ready canola is that the Roundup-resistant gene might jump to conventional canola or closely related weed species, or that seeds of the Roundup Ready variety can themselves become volunteer weeds by blowing from farm fields, trucks, or storage bins across the provinces. These are potentially serious side effects of GM crops whose impact is still unclear, but conventional producers are unconcerned. From their perspective, the current practice of rotating herbicide use in fields, roadsides, and right-of-ways can keep the resistant weed types from becoming established, at least in habitats of concern to humans.

Consumer fears about the safety of GM foods also do not resonate with producers. Farmers do not agree with the public outcry against GM foods, believing that the regulatory system is sufficiently rigorous to catch any problems. Jonothon Roskos voices a typical view: "I would say everything fits my comfort zone. I have a great deal of faith in our regulatory system." So does Ray Wilfing, a highly respected Meadow Lake, Saskatchewan, farmer: "I think consumers have had pretty good food lately and they just don't understand what's involved. There's been a lot of fear-mongering, too, and people don't understand what's been going on, I guess. Yes, consumers are concerned, but they're concerned about everything."

▉▉▉ My prairie foray made it clear to me that the dominant view on the farm is that GM canola is a safe and cost-effective crop, and that there are so many sources of GM seed that no firm can dominate the market. Even the minority of farmers opposed to GM crops and driven to vehement protests and lawsuits against companies like Monsanto are motivated more by concerns about the declining farm system than by panic over environmental or food issues involving biotechnology.

In this, farmers differ from the urbanites who buy the food they produce. The latter worry about the safety of genetically engineered foods, but farmers are not concerned about the health risks of GM canola in food, and trust government regulators. An earlier generation of regulators caught potential problems with rape oil for humans and livestock, and canola was the result. Today's regulators have examined GM canola varieties, and no obvious health problems have emerged. Ironically, consumers sympathize with the plight of farmers and hold the family farm in high esteem even as they express alarm about new scientific developments that could save Canada's family farms.

The current science of GM crops is only in its infancy, and most producers think that if there is a reason to be optimistic about the future, it lies in the competitive edge they might achieve through the new technologies and crops that are under development. Canola varieties already are available that confer agronomic benefits such as herbicide tolerance and pest resistance, and laboratories all over the world are competing to create the next generation of canolas that will produce more specialized proteins and lipids for industrial, pharmaceutical, and nutritional uses.

New transgenic varieties of canolas can now be created in the laboratory within three to four months, and these crops will be tested, regulated, and planted in farmers' fields within two to five years. Included in the cornucopia of novel GM canolas under development are seeds containing such obviously beneficial substances as anticoagulant proteins for medical use; edible vaccines and antibodies against cholera, diarrheal diseases, measles, and the bacteria that cause dental cavities; biodegradable plastics; clean-burning biofuels and novel industrial oils; and plants useful for bioremediation, such as the uptaking and degrading of chlorinated solvents deposited in waste dumps or spread by accidental spills.

There is little about these advances that concerns canola producers. Rather, the dominant sentiment among farmers, whether they grow canola or other commodities, has always been to do whatever it takes to produce a crop. Canola farmers today trust that biotechnology will provide safe solutions that will preserve their farms by inventing new varieties derived from contemporary genetic engineering techniques, just as the creation of canola helped maintain economically viable farms for their fathers and grandfathers.

Growers around the world are facing challenging economic hardships, trapped somewhere between rising prices for input costs and increasing yields that ironically lead to falling prices for their crops. For most farmers, Schmeiser's issues about technical use agreements, patent rights, and multinational corporate politics pale beside the pragmatic concern of producing a crop whose bottom line will not be red. Producers view the environmental and food safety risks of GM crops as small to nonexistent, and see diversification of GM products as the only viable way to increase farm income and profits.

Farmers have not lost faith in technological advances. Rather, they believe that yesterday's technologies must be improved

upon in every generation if they are to maintain their competitive edge. The farm view is simple: if progress through genetically modified crops will keep farmers and their children working
on the farm, then we should embrace the technology and get the
crop in the ground.

It just doesn't make sense to risk all that has been achieved by organic growers for a technology that may not work, that will not benefit local tree fruit growers, and that consumers are rejecting in ever increasing numbers.

Linda Edwards, organic farmer, addressing the Economic Development Committee of the Okanagan-Similkameen Regional District, 1 June 2000

Saving the Bugs

Wenatchee, Washington, bills itself as "The Apple Capital of the World," and this theme is carried out in the names of the Apple Shoppe Café, the Apple City Snowmobile Club, Applebee's Grill and Bar, and of course the requisite small town baseball team, the Apple Sox. Spring is marked by the annual Washington State Apple Blossom Festival, fall by Apple Days, and winter by the Apple Cup Ski Event.

It is in late summer, though, that the apple motif is most prominent, for that is harvest time. In the Wenatchee orchards, straight rows of apple trees and stacked red bins overflowing with freshly picked fruit form an almost unbroken vista along every creek and river, bordered by the dry desert terrain characteristic of the region before the cultivation of fruit trees.

Apples have no natural business in this arid land of flat tabletop mesas, scree slopes, and sagebrush. Apples evolved elsewhere, in the deep ravines and forested slopes of Kazakhstan, an area that is culturally, biologically, and environmentally worlds apart from

the dry eastern foothills of the Pacific Northwest's coastal mountains. Yet human ingenuity, in the form of massive hydroelectric
dam projects, extensive irrigation systems, and modern transport
systems, has transformed this landscape into apple country.

These central Washington orchards extend far north into
Canada and far south through Oregon, but they are not as picturesque as they appear in carefully cropped publicity postcards.
The impact of humans is obvious throughout the region, from
the networks of power lines and irrigation sprinklers, to the small
dumps of abandoned farm equipment that mar the uncultivated
landscape, to the dams, small and large, many bearing signs
proudly proclaiming that these tamed water systems are "owned
by the people we serve."

But it is the less visible chemical impact that may be the most
invasive: most of these orchards are maintained by ecologically
questionable applications of insecticides, fungicides, herbicides,
and artificial fertilizers. Not all growers have accepted dependence on chemicals, however. Organic farming has become increasingly prominent in the region, as it has throughout the
farming world for many commodities. The organic fruit industry
has blossomed in recent years, driven by market demand, public
concerns about food contaminants, increasingly sophisticated organic cultivation methods, and higher profit margins for organic
growers than for conventional farmers.

Organic farming has become a viable alternative to the high-
input and low-profit agriculture that has come to characterize
most of the developed world's food-production systems. Organics are poised for continued growth, but another management
tool has come onto the scene that threatens the survival of organic farming, genetically modified crops.

Biotechnology provides both philosophical and practical challenges for organic growers. Genetic modification raises thought-
provoking issues about just how far we can select and tinker with

crops before they cross the boundary into the nonorganic realm, and broader issues about the relative sustainability and productivity of high-input conventional agriculture and restricted-input organic farming. But more practically, the key questions are clear and direct: Will wind-blown or bee-carried pollen from GM crops contaminate organically grown crops, and will consumers tolerate the infiltration of genetically modified genes into organic produce?

In 2000 I spent a few late August days traveling through central Washington, visiting organic producers, vendors of organic farming supplies, and brokers for organic produce. It quickly became clear that the stereotypical hippie organic farmer with an economically marginal lifestyle is now a myth; these growers are efficient businesspeople who run profitable organic farming enterprises. Organic farmers are now comfortable with being described as capitalists.

Harold Ostenson is one organic grower who has made the transition from marginal to successful. He bought his first orchard in 1976, at Frenchman Hills, Washington, following a stint as a naval engineer in Vietnam. Ostenson is in his mid-fifties, with a face that has seen some sun in its time, and he talks about farming with a slow, measured intensity. He described the typical organic farmers of the 1970s as "outcasts, nature lovers and tree huggers, people marching to a different drummer. The organic grower probably forgot to get a haircut for a few months. In the sixties and seventies, the early core was the hippie with a quarter acre of tomatoes who was saving the bugs. You went to the Ma and Pa grocery store and at the end of the aisle was a box of fruit that had all these bug holes and a little sign that said organic."

No longer. Ostenson's current business typifies what has happened to organic farming, which has grown from a string-and-baling wire backyard craft to a sophisticated international industry that has become a mainstream component of contemporary agriculture. After slowly converting his first orchard from conventional to organic crops, and hand-packing his first fruit in an abandoned potato shed, Ostenson has transformed his farm into a complex assembly-line operation that packs organic nectarines, cherries, apricots, apples, peaches, pears, and potatoes grown on his own extensive acreage as well as on many neighboring organic farms. Some of the local organic farms have grown as large as a thousand acres in size, many are owned by absentee landlord corporations, and the fruit he packs is brokered through a San Francisco company that ships organic produce all over the world.

In short, organic agriculture is an industry growing with the speed and confidence of another rapid-fire contemporary corporate success story, e-commerce. Organic food sales grew at a rate of 20 to 25 percent per year throughout the 1990s, compared with a rate of about 4 percent for conventional food, and that pace is continuing. Sales reached $6.6 billion in the United States in the year 2000, making up slightly more than 2 percent of the food market, and 6,600 U.S. farmers were involved in growing these crops. Downstream, companies selling organic foods also have grown dramatically. Nature's Path, for example, is the largest seller of organic cereals in North America, with annual sales of $70 million and 150 employees. Perhaps organic corporations are not yet Kellogg's or General Foods, but they certainly are no longer Ma and Pa companies.

There are numerous and related reasons why the organic industry has had such a steep growth spurt, some not specific to organic agriculture. For one thing, transportation has improved during the last few decades to a point where rapid distribution of fresh produce limits deterioration in the quality of fruits and vegetables while in transit, and reduces spoilage. Another factor has been prosperity, with the voracious American consuming public more able to pay the slightly higher prices that often are charged for organic food. Consumer concerns about pesticides in food and high-profile toxic residue scares such as that involving Alar in apples haven't hurt either.

But the most significant factor driving the success of organic farming has been economics. The costs of high-input conventional farming have risen dramatically, especially the prices of synthetic pesticides and fertilizers; and increased production from mega-farm conglomerates has driven the price of conventionally grown crops down, all of which has made it difficult for conventional farmers to reap a profit. Harold Ostenson sees this situation as clearly favoring the organic farmer: "Organics are successful because of the economics. The conventional guy is up against the wall. His costs are going up and his price is going down, down. He's caught in that trap. That's the wall I don't see on the organic side . . . The trade-off has always been based on the economical aspects—I can do more for less—and that's what the chemical thing is all about. That's not true anymore; organic growers spend equal or less money to grow fruit compared to conventional growers, and that's a whole different thing."

Organic farming has changed, and although consumer perceptions may perpetuate its image as a fringe activity, the reality is different. As Harold Ostenson put it: "For a long time the organic thing was a way of living—'I'm living cleaner, I'm not polluting'—as opposed to 'this is an economically viable production business.' On the left and right of me I had the Mother Earth concept, but my thrust was more how could you commercially, economically, viably have a business growing fruit. I guess I was a little less into the religion of just being Joe Organic."

Given the economic powerhouse that the organic industry has become, its universal opposition to genetically modified crops is not surprising. There now is a multibillion dollar industry at risk from biotechnology, and perceived self-interest is an important component of the industry's aversion to genetic modification. Some would describe the issue more strongly as one of self-preservation, especially following a 1998 recall of tortilla chips because of contamination from genetically modified crops.

The culprit was pollen from genetically modified corn that was blown into a nearby field of organic corn. Apparently, a small corner of the organic field was contaminated by this cross-pollination. The harvest from the entire field went to a small Wisconsin company called Terra Prima, which processed the corn into organic tortilla chips and shipped them off to Europe under their brand name, Apache. A German consumer magazine happened to run some random DNA analyses on organic products as part of its research for a story it was writing on GM crops, and rang the alarm bells when DNA from GM corn popped up in the tortilla chips. Although less than 1 percent of the corn used in the chips was genetically modified, Terra Prima pulled back the entire batch of 87,000 packages, and kept its product off the shelves for four months, at a cost of about $250,000 in lost sales. The company now tests every batch of organic food that it purchases for potential GM contamination, at considerable cost, in order to restore and maintain consumer confidence in its organic label.

All organic products are similarly vulnerable in the marketplace because they are not necessary. Conventionally grown food is readily available, at a lower price; organic farming is viable only because shoppers choose to buy organic products. Consumers may shop for organic products for many reasons, but at the core is an underlying perception that organically grown food is healthier for people and friendlier to the environment than food produced by conventional agriculture. Thus any trace of contamination is taken seriously by growers, and reliable certification of food as being organic is an essential regulatory component of the industry.

GM crops present particular issues for certification, as the Terra Prima incident proved, but certification issues are not new to organic farmers. Even defining "organic" is troublesome. Definitions touted by farmers, packers, government regulatory agencies, distributors, and stores abound, each with its own subtle twist. The terms found in organic definitions tend toward the user-friendly but vague, including holistic, sustainable, harmonious with the environment, and ecologically balanced. Fundamentally, organic practices exclude the use of antibiotics, hormones, and synthetic pesticides and fertilizers; favor the use of biological and physical methods to manage pests; and apply animal and green manures to fields. Organic farmers use the natural ecological processes of soil organisms, nutrient recycling, and species distribution and competition as their principal farm management tools.

Producers and consumers alike seem to think they know what "organic" is, but the fine print in organic certification schemes can run to hundreds of pages. A reasonable summary of what organic certification means might be: Organic farmers are purists about what happens on their own farms but realists about contaminants that result from the activities of their neighbors. Certification regulations provide for zero tolerance of nonorganic methods in on-farm practices, but commonly allow some wiggle room, although limited, for problems caused by drift. Synthetic chemical pesticides are a good example. Organic certification would be denied if a farmer sprayed any banned substance, but U.S. regulations allow organic food to have 5 percent of the acceptable EPA pesticide residue level for conventional crops if those residues originated from a neighbor's pesticide spray.

The organic industry is attempting to grapple with genetically modified crops in the same way they deal with synthetic pesticides, by banning their use on farms. The current American Organic Standards state clearly that "genetically engineered/modified organisms . . . or products produced by or through the use of such organisms are not compatible with the principles of organic production." However, organic certifiers are beginning to consider accepting some level of GM contamination in recognition of the difficulty organic farmers face in producing GM-free products.

A report from the Genetically Modified Organism Task Force (GMO Task Force) of the U.S. Organic Trade Association highlights some of these issues. As expected, it supports banning the use of GM crops in organic farming, but it had more trouble with GM residues such as those found in the Terra Prima tortilla chips. The task force's report expresses frustration at the organic industry's lack of power to prevent conventional farmers from using GM crops, and suggests that "unintentional, low-level GMO contamination should not be cause for decertification of a product." It also politely footnotes that statement with a delicately understated qualification: "The Task Force is not unanimous on this point."

A close reading of this report reveals just how widespread the impact of GM crops may be on organic agriculture, and how important some degree of flexibility about the contamination issue may be for the survival of organic farming. Cross-pollination with pollen from GM plants that blow into or are carried by bees into organic fields is one obvious danger, but there are many other avenues by which the products of genetically modified crops could make their way into organic food.

Split operations and transitional farms are one source of contamination. Many organic growers also run side-by-side conventional operations, and are able to keep them separate with careful management, because slight contamination is acceptable under the standards for organic products. With GM crops, this practice may no longer be acceptable if consumers demand zero toler-ance for GM residues. Other farms are defined by the American Organic Standards as "in transition" during a three-year change-over period from conventional to organic farming. This length of time may be sufficient to clear farms of conventional pesticide and fertilizer residues, but it may not be long enough to rid fields of volunteer weeds remaining from an earlier GM planting.

Inputs and processing pose other problems in the quest for purity. Organic production is no different from conventional farming in requiring the application of fertilizers, minerals, and compounds to adjust soil acidity. Livestock require feed, vitamins, amino acids, glucose, and vaccines. All of these inputs are poten-tial sources of GM contamination, and would have to be carefully sourced and certified. The processing of organic foods is another problem area. The task force suggested that "organic processors must intensify efforts to find non-GMO sources for minor ingre-dients, to ensure that there is no use of materials that may even inadvertently contain GMO's."

The seriousness with which the task force viewed these issues is apparent in its recommendations for a labyrinthine and com-plex paper trail to document the lack of any trace of GMO's in organic food. It suggested legally binding affidavits signed by quality-assurance managers, third-party reviews to ensure that certification rules have been met, and national lists of approved materials. Even these layers of paperwork won't assure purity, so the task force recommended that producers and packers avoid la-beling organic food "GMO free" or "No GMO's," and use terms like "No GMO Ingredients Used" to allow for the possibility of contamination from outside sources.

The crux of the issue is that people who buy organic foods tend to be purists, and may not accept any level of GM contami-nation. When the U.S. Department of Agriculture asked for com-ments about proposed organic certification rules, it received over

275,000 replies that nearly universally opposed the use of GM technology in organic agriculture or in the production of minor ingredients used in growing or processing organic food. The Terra Prima tortilla chip incident also indicated that zero tolerance may be the standard consumers impose on organic foods. The world trade markets are debating a proposal to allow up to 1 percent contamination of conventionally grown food by GM varieties, but this may not be stringent enough for consumers of organic products. As the GMO Task Force report points out, "Exporters are finding that products showing contamination by GMOs may be rejected at any level determined by the customer, regardless of government standards."

If the organic marketplace continues to speak in that voice of zero tolerance, organic producers may not be able to achieve what their customers are demanding. It will become increasingly difficult for organic growers to produce GM-free products as biotechnology in conventional farming continues to expand. The options are limited: either rein in agricultural biotechnology or compromise on consumers' stipulations for GM-free organic food.

Buyers of organic food may have little room in their value systems for compromise, especially those in the strict European markets. Organic farmers soon will be trapped between the proverbial rock and hard place, stuck between an inflexible public and biotechnological trends over which they have no control. The future of organic agriculture will be murky until this difficult conundrum can be resolved.

■■■ Besides worrying about marketplace reactions, organic growers are concerned about the more direct impacts of biotechnology on production systems. The most significant of these threats to the future of organic farming are the development of pest resistance to the organically acceptable biopesticide Bt and disruptions to the predator and parasite populations that control many pests on organic farms.

One of the common misconceptions about organic agriculture is that growers simply plant crops and let nature take its course. Not so. Organic farming requires considerable input to succeed, although cultivation methods and the products used for fertilizer, soil conditioning, pest control, and disease management differ substantially from those employed in conventional agriculture. An organic farm may superficially resemble a conventional farm, but a closer look reveals a complex array of techniques and over two hundred approved products that are very different from conventional management tools.

Organic apples, for example, require more labor than conventionally grown apples, and a cornucopia of natural products is needed to produce high-quality fruit. Differences in organic and conventional apple growing fall into four broad areas. First, organic farmers control weeds by tilling the soil rather than by spraying herbicides, doing considerably more work but eliminating the use of chemical herbicides. Second, although all commercial apple growers use thinning methods to reduce the number of fruit produced by each tree, thereby increasing the size and quality of each fruit, conventional growers use chemical sprays to thin the fruit shortly after the trees flower, whereas organic growers thin their apples by hand, again expending more labor but reducing on-farm chemical use. Third, conventional apple farmers apply synthetic nitrogen, phosphorous, and other nutrients to the soil, whereas organic growers use more complex composts and manures to build, condition, and maintain soil quality.

The most significant differences relevant for genetic modification issues are in the fourth area, pest management, where organic growers have substituted predators, parasites, and bio-

logically based viral and bacterial pesticides for the synthetic
chemical pesticides that characterize conventional apple cultivation. Both conventional and organic apples are besieged by an array of native and imported pests. Cultivated apples and the trees
that bear them are as tasty to insects as their fruit is to us: aphids
and scale insects pierce leaves and feed on sap; cutworms, leafrollers, winter moths, and skeletonizers chew on leaves; and fruit-
worms, mullein bugs, and codling moths devour the fruit.

The biggest fear about the potential side effects of GM crops
on organic production systems is that they could induce resistance to the widely used bacterial pesticide Bt in the numerous
moths and caterpillars that feed on crops. Organic orchardists are
afraid that conventional apple growers might be tempted to begin using varieties bioengineered with Bt genes. If so, pests
would be continually exposed to the toxic Bt proteins expressed
in the perennial orchard trees, rather than being briefly exposed
during the application of sprayed formulations of Bt, which
break down quickly in the environment. The resistant pest insects would soon spread from GM orchards to organic farms, an
obvious problem for organic producers who rely on Bt as their
primary tool for controlling pestiferous moths and caterpillars.

More broadly, the use of Bt varieties of any crop might induce
resistance in apple pests. This may seem far-fetched, given that
most apple pests don't feed on the crops, such as corn, soybeans,
cotton, and canola, that currently have Bt varieties, but there may
be some substance to this anxiety. Bt is a soil-dwelling bacterium,
and recent studies have suggested that the proteins expressed in
Bt varieties can migrate into the soil and persist for many years,
potentially spreading to nearby fields and orchards. If so, the result would be a persistent, low-level expression of an insect-toxic
material throughout the environment, a textbook example of
one mechanism by which resistance can be induced in pests.

Organic farmers are deeply concerned about the induction of

resistance, because sprayed formulations of Bt are by far the most important and economically successful pesticide option approved for organic farming. There are others, but they tend to be fringe products such as diatomaceous earth, garlic barrier, hot pepper wax, insecticidal soaps, and lime sulfur, helpful in a few situations but not nearly as widely used as Bt.

Bt is a dream product for organic growers because it does not affect the predator and parasite complexes that they rely on to substitute for synthetic chemical pesticides. Conventional insecticides have many disadvantages for organic production, but the most practical one is that they work over a broad spectrum, and kill beneficial as well as harmful insects. Bt, in contrast, is highly specific to the larvae of moth and butterfly pests, and has no effects on the natural or introduced beneficial insects that make up an important component of pest control in organic systems. Further, Bt has shown no effects on human health after almost forty years of use, and is therefore as close to being entirely safe as any pesticide ever studied.

Organic growers also are worried that genetically modified crops will induce reductions in naturally occurring or released predators and parasites, but this fear may be driven more by their visceral opposition to anything GM rather than by an actual threat. Organic growers rely on wild or purchased and released biological control species such as ladybugs, predacious mites, parasitic wasps, and lacewings to manage many of their pest problems. Since many GM varieties are designed to reduce pesticide use, GM crops could be considered to be beneficial in that their aimed-for reductions in pesticide use should favor biological control agents.

Organic producers have not bought this argument, but rather focus on as-yet unproven food chain effects whereby parasites and predators would be harmed by feeding on pests which in turn had fed on genetically modified crops. A few laboratory studies have indicated that this could occur, but field analyses have not found any reductions in beneficial organisms induced by the use of GM crops. Nevertheless, organic farmers have a deeply ingrained distrust of the industries supporting conventional farming, and assume that it's only a matter of time before harmful effects of GM crops on biological control agents are discovered.

■■■ The contrasting nature of traditional crop selection and contemporary GM techniques provides another focal point for opposition from the organic industry. For conventional growers, GM apples are as tempting today as the Garden of Eden variety was for the biblical Eve, because transgenic orchards have the potential to reduce the costs of pest and disease control and to improve the quality of the fruit. Conventional growers of all crops tend to view genetic modification pragmatically, as a cost-cutting and product-enhancing tool, whereas organic farmers view biotechnology as a source of threats to their growing methods and, more philosophically, as an ethically dubious endeavor.

Apples are an interesting crop to focus on when we ponder the history and methods of artificial selection for crop varieties and consider whether biotechnology represents a minor modification to traditional plant-selection methods or a quantum horticultural leap in how we humans mold the biological world around us for our own purposes. The contemporary domesticated apple, *Malus domestica,* has a fascinating and complex history of transgenic mingling between closely and distantly related plant groups, initially exchanges between different types of wild plants and later exchanges brought about by grafts and crosses between increasingly tamed varieties.

Apples are in the family Rosaceae, which includes the namesake roses but also pears, plums, peaches, strawberries, and cherries. It is a plant family whose genes and chromosomes exchange easily between species. Apples are thought to have originated when a primitive plum cross-pollinated meadowsweet, another plant from the Rosaceae family. The apple descendant wound up with eight chromosomes from plum and nine from meadowsweet, and eventually diversified into about thirty-five species with a natural distribution that stretched across central and northern Asia into Canada. Most of their fruits were tiny, bitter, and toxic, with seeds containing high concentrations of cyanide to deter pests from eating them.

The origins of cultivated apples are debatable, but most experts have singled out southwest Asia, near the Caucasus Mountains, or south-central Asia, near the Chinese-Kazakhstan border. Both areas are hubs for ancient trade routes, by which the earliest cultivated varieties made their way to Europe and eventually around the globe. Close to a dozen apple species inhabit this region, and their fruits can be green, yellow, red, or orange in color, large or small in size, bitter and inedible to sweet in taste. These species have been mixed by natural pollination and selected by plant breeders over many millennia to a point where the origins of specific traits are difficult to determine. What remains is the one domesticated species, with a wide range of highly selected traits, and a few remnant pockets of the original ancestor apples in deep ravines and valleys far from human habitation.

The diverse origin of apples, and the unusual ability for apples of different species to mingle their genes, have made them among the most naturally transgenic of species. Their propensity for mixing up genetic diversity has allowed plant breeders using traditional techniques to create hundreds of novel apple varieties, and to select useful traits to incorporate into standard apple varieties. For example, characteristics such as cold-hardiness and resistance to rots, blights, scabs, fungi, and bacteria all have been bred into commercial apples. These traits were initially mingled in by careful transfers of selected pollens, and desirable offspring were propagated by grafting their buds onto standard rootstock bases.

This process can take decades, but today plant breeders can use biotechnology to accomplish gene transfers within a single apple generation. It is not difficult to understand why these easily accomplished genetic modifications would tempt conventional growers to put aside ethical concerns in favor of expediency. Their management systems are dependent on specific chemical inputs to control pests and diseases, and it is both conceptually and practically simpler to continue with that strategy through bioengineering rather than to switch to an organic production system.

GM apple varieties have been under development for many years, but so far opposition from consumers and organic growers has prevented transgenic apples from expanding into commercial production. Field tests have been conducted in the United States for trees that are resistant to apple scab, fire blight, and lepidopteran pests such as codling moth and leafrollers. Other varieties have been developed with fruit-enhancing characteristics added, including reduced ethylene synthesis to retard spoilage during storage and increased sugar alcohol levels to improve taste. But even these field tests have been controversial. In my own province of British Columbia, organic growers from the Okanagan Valley successfully lobbied the provincial Ministry of Agriculture to prevent testing of a browning-resistant apple developed in the laboratory by a private company in partnership with the federal-level Agriculture Canada.

Organic apple growers view bioengineered fruit quite differently from the way conventional farmers and the biotechnology industry do. One concern is market driven: that cross-pollination between transgenic and organic apples would result in fruit that was unacceptable to the GM-averse buyers of organic food. That is important, but the position of organic farmers on GM issues also arises from a genuine and visceral abhorrence of the philosophy behind genetic modification.

Traditional selection methods do pose a few ethical issues, especially concerns about our human role in reducing natural variation by selecting for particular traits in domesticated apples. But the contemporary transgenic methods by which breeders quickly and precisely mingle genes from bacteria, animals, and other plants into apples pose much more challenging questions about which crop-selection techniques are appropriate.

The essence of an acceptable apple from the viewpoint of organic proponents is a fruit that was developed through a combination of selection in nature and gentle breeding by orchardists. The more forceful intervention embodied by genetic engineering feels unnatural to them, as out of place as transplanting a head of cabbage onto a baboon's body would seem to most of us. Some might say they are overreacting, but organic growers have intense and deep-seated opinions about evolution and selection that underlie their strong aversion to biotechnology.

I spoke with Phil Unterschuetz and Mariah Cornwoman from Integrated Fertility Management (IFM) in Wenatchee, Washington, about mixing up genes. Their company sells products and services for organic agriculture, ranging from organic fertilizers and pest-control products to soil and plant analytical services. IFM's dilapidated offices and warehouse literally groan from the weight of products that move in and out quickly during the growing season, and the old wooden floors have collapsed and been repaired many times. The boom in organic farming has not passed IFM by, however, and the company is about to move into a larger and sturdier facility.

Unterschuetz is easygoing, thin, and fit. Initially, he regarded his participation in the organic supply industry as merely a business venture, but he soon became a strong advocate of the organic way of life. Mariah Cornwoman does a lot of the warehouse work and has the muscled appearance that comes from hefting endless bags of fertilizer; she embraced the organic view after becoming disenchanted with conventional, chemically intense farming. They both are articulate about the philosophy underlying the organic approach, and both are adamant about letting nature take its course with minimal human intervention.

Unterschuetz's support of organic farming is based in part on its low impact on the genetic integrity of the natural world. His concerns about mingling genes between distant organisms focus on the unknown consequences for nature of transgenic crops: "What does organic mean? It somehow has a meaning to most people that means natural. There's no chance of that if a crop has a planaria gene in it, or something like that. I'm afraid we're going to substitute a set of high-tech solutions that may have implications for the rest of life on this planet. We don't have a clue about where those genes are going to go and what will result from it."

Cornwoman's opposition to GM crops grew out of respect for what evolution has produced, and her belief that we humans cannot and should not try to improve on that: "GMO's certainly increase our genetic base, but I don't think in a positive way. It's like thinking that we can circumvent millions of years of evolutionary natural selection with one little splice. I don't think you're going to outdo evolution in a few years of genetic breeding."

■■■ The opposition to biotechnology from the producers, distributors, and consumers of organic food springs from a well of deeply ingrained beliefs, much more powerful motivators than the practical issues of market share, threats to the integrity of organic production methods, and selection methods for crop varieties. The organic industry may have gone corporate, but it has retained a genuinely alternative perspective on farming, one that is profoundly different from that of typical agriculture and that has been the source of its resistance to biotechnology.

The philosophical divide between conventional and organic farmers is obvious from the many quotations I saw pinned to bulletin boards and taped to walls in the offices of organic industries, such as "Changing the world, one fruit at a time," and "We all live downwind." The books on their office shelves reinforce the theme of concern for the environment: *Nontoxic, Natural, and Earthwise; The Real Dirt;* and *The Natural Enemies Handbook.*

It is at the intersection of farming and environmental awareness that the conflict between proponents of genetically modified crops and practitioners of organic agriculture comes into clear focus. Organic producers view their systems as enhancing the environment, reducing waste, and allowing natural ecological cycles to function within and outside agricultural fields. Minimizing impact and treading lightly on the earth are as fundamental to organic growing as producing pesticide-free food. Appreciation for nature and its infinite intricacy are central to the industry's opposition to biotechnology. As Phil Unterschuetz put it, "I carry a kind of awe, a sense of reverence for the complexity. I think it's just utter hubris to think we can not only understand it but manipulate everything around us for our own pleasure. All of the evidence I see leads me to believe that everything we do on this planet reduces speciation, reduces complexity, reduces diversity, and simplifies, simplifies, simplifies the ecology of the world, and that is not sustainable."

Organic producers also fear unleashing new technologies that they do not think have been fully tested to rule out adverse ecological effects. From their perspective, organic farming methods work just fine and produce crops economically, and there is no agricultural problem for which genetically modified crops provide a necessary solution. Harold Ostenson told me that "it's not like there's a huge meteor that's going to directly hit the earth, and we've got one year to figure out the solution or we're all going to get hit. Once you let the genie out, you can't get it back inside the bottle. Once you make your mistake, once you pollute the environment, you're not able to go back without years and years of trying to get it out of the system, and you might never be able to."

Organic farmers also object to the attitude adopted by proponents of GM crops. Organic growers have developed sustainable agricultural systems that improve the environment and generate jobs and income for thousands of farmers and companies that distribute and sell organic food. They want some respect, both for the validity of their methods and for their lifestyle and their views about the environment.

Greg Kahn has run Cascadian Farm in the Pacific Northwest for thirty years, and he's proud of his high-quality products and upset about how biotechnology advocates and conventional farmers demean the organic way of life: "What the GMO supporters are ignoring is that organic is a lifestyle decision. They call us scientific illiterates. Their simplistic view of our opposition pisses me off. You wouldn't argue with someone who observes religious dietary laws; it's a choice that people make. Gene tinkering flies in the face of what some people think is a natural law. It's not only because we think it's a hazard—we believe this is a transgression of the natural world."

The fears of organic farmers are not unfounded. In fact, GM crops do threaten organic agriculture in practical ways. Most notably, the mingling of GM genes with organic food seems inevitable, and while the extent of contamination may be low, tolerance by consumers may be zero. Also, the overuse and misuse of Bt varieties of GM crops by conventional farmers is likely, and the resulting induction of pest resistance would devastate organic growers.

Might conventional and organic farming survive side by side? Compromise would be needed on the part of consumers of organic food, especially in accepting a small amount of GM contamination. Buyers of organic food show no indications of doing this willingly, but perhaps if the issue was forced by the stark choice of slightly contaminated organic food versus no organic products, consumers might come around.

Conventional growers would need to be sensitive to the concerns of neighboring organic growers, and limit their acreage by agreeing to wide barrier zones between conventional and organic farms, up to five to ten miles for some crops if the tolerance level for pollen transfer is zero. Perhaps entire farming districts would need to be delineated for conventional and organic farms, separated by a broad nonagricultural barrier zone. This might be feasible if mandated by law, but it is unlikely to come about otherwise.

These compromises would resolve some of the practical problems, although they would not be popular and could easily prove to be unworkable. But practical concessions do not address the fundamental ethical differences between proponents of conventional and organic agriculture.

The philosophical divide looms large, a crevasse of differing perspectives and lifestyles that will be difficult to bridge. The approaches of conventional producers to farming may be internally consistent, as are the methods used by organic growers, but there seems to be little overlap in their attitudes about the world, and few areas in which conciliation could span the gap.

For all these reasons, it will be extremely difficult to reach a compromise between the two parties. Nevertheless, there is too much at stake to stop discussing the ways of achieving a rapprochement. At issue is the survival of organic farming, and the philosophy that undergirds it has too much to offer all of us to allow it to founder on the rock of biotechnology.

Anything under the Sun

John Doll is head of the Biotechnology Examination Group of
the U.S. Patent and Trademark Office, and he is passionate about
patents. Applicants who enter his department expect the confusing regulations, excessive paper-pushing, and maddening bureaucratic roadblocks typical of government offices. Instead, they are
surprised by the enthusiasm, clarity, and get-it-done attitude he
and his staff bring to their consideration of applications for
patents concerning genetic engineering.

Doll's department is responsible for patents for biotechnology
and pharmaceutical products, and it has had enormous influence
over the progress and profitability of these industries. The stakes
are high and the lobbying is intense, because hundreds of millions of dollars in corporate profits depend on the acquisition of
patents to protect inventions from marauding competitors. Fortunately for these industries, Doll's group is efficient, and strives
to facilitate and accelerate rather than prevent or delay the issuing
of patents.

The exclusive right to intellectual property bestowed by Doll's
group has been a cornerstone of the biotechnology revolution.

The science of genetic engineering has been fascinating, and the uses to which biotechnology are being put are stunningly inventive and functionally impressive, but genetically modified organisms would be only an interesting academic sideline if there was not money to be made. The heavy investments in research that have driven corporate biotechnology would not have been forthcoming without the product protection provided by patents.

The granting of patent protection for biotechnology products makes perfect corporate sense, but it has been controversial. Legal issues about proprietary claims on life itself have challenged the decisions by the Patent Office to award biotechnology patents. These judicial assaults have overlapped with ethical questions about ownership of the world's genetic resources and moral questions about the appropriateness of genetic engineering. Both the ritualized procedures of legal action through the courts and the more free-wheeling tactics of lobbyists have been used by protesters to contest the rationale underlying genetic engineering.

These conflicts have taken place internationally but reached their statutory apogee in the United States, where patent protection most heavily favors inventors, the potential for corporate profit is at its greatest, and litigation is often the first rather than the last recourse. The U.S. Patent and Trademark Office (USPTO) has navigated through this potential legal and ethical minefield with an unusually clear focus on its primary goal. It has defined its mandate simply, and kept its fundamental objective of granting patents front and center through intense protests and repeated legal attacks.

Doll explained the unswerving philosophy that has guided the patent office through the biotechnology wars: "It's our job to help an applicant get a patent. Gene companies would not be doing the research and making the discoveries if there was not a potential payoff. We either allow people to recoup their research and development and make a possible profit for their shareholders or we simply won't have the technology."

The concept of patent protection may have been honed through the business-boosting policies promoted by the U.S. government, but patenting began in Europe when sovereigns began to grant privileges. The word "patent" itself means "to open" in Latin, and refers to the waxed seal on documents called "letters patent" that confirmed sovereign decrees. The first patent describing an invention was granted in the Republic of Florence in 1421 for a barge with hoisting gear to transport marble, and the idea of patent protection soon spread throughout continental Europe and England, and eventually to the New World. The important role that patents would play in protecting intellectual property was understood by the framers of the U.S. Constitution, who considered it significant enough to enshrine patent protection in Article I, Section 8: "Congress shall have power to promote the progress of science and useful arts by securing for limited times to authors and inventors the exclusive right to their respective writings and discoveries."

The first U.S. patent was granted in 1790 by Thomas Jefferson, himself a noted inventor and then Secretary of State, for a method of manufacturing potash. Since then, over six million patents have been granted in the United States, many of them eventually winding up the object of complicated lawsuits, not unexpected when profit and pride of invention are at stake.

The U.S. Congress passed the Patent Act in 1952 to clarify what could be patented and provide a benchmark definition with which to resolve the innumerable lawsuits brought in the U.S. courts. The act defined patentable material simply, as "anything under the sun made by man," and that is the standard still applied today. That concept is relatively easy to interpret when a gadget is in question, but genetic engineering has challenged the patent office and the courts to decide what has been forged by nature and what has been made by man.

I met with John Doll at his Crystal City, Virginia, office to discuss the ins and outs of patent law and philosophy. It was dress-down Friday, a day well suited to his gregarious but intense nature. Doll is trim, fit, and somewhere close to fifty in age. His office walls are plastered with family pictures and awards, and lined with shelves groaning under stacks of applications and computer printouts. His academic background was in physical chemistry, and although his career as a patent examiner began in that field he quickly retrained to focus on proteins and genetic engineering. His motivation was simple; government red tape and overly cautious interpretations of patent law were impeding the growth of young biotechnology companies, and his self-imposed mission at the U.S. Patent and Trademark Office was to change that.

The guiding principle behind all patents is not to grant the right to make or use an invention, but rather to exclude others from making, using, selling, or importing it. In return for twenty years of exclusion, the public gets eventual access to full information about the invention. As Doll put it, "The patent office is an administrative agency that allows somebody to exclude somebody else. We give monopolies. But the monopoly's only for a limited amount of time, and what we get for giving people that exclusive right is the complete disclosure of how to make and use their invention, a complete enabling disclosure."

Patents protect intellectual property, and literature from the patent office describes intellectual property with a mixture of sentiment and pragmatism: "It is imagination made real. It is the ownership of dreams, an idea, an improvement, an emotion that we can touch, see, hear, and feel. It is an asset just like your home, your car, or your bank account." The standards for granting a patent are conceptually simple, requiring that an invention be novel, useful, nonobvious, and described sufficiently in an application so that anyone can construct and use the invention when the patent has expired.

Doll refers to his group as an "art unit," as in state-of-the art, and he employs fifteen examiners to adjudicate the thousand or so applications for patents involving transgenic plants that are being considered at any one time. Patenting organisms may seem odd, given the statutory restrictions that U.S. law places on products-of-nature patents. However, in 1930 the U.S. Congress passed the Plant Protection Act, which extended patent protection to plant varieties, and the courts have upheld this legislation by ruling that new plant varieties developed through research and human-directed selection were invented by man rather than evolved by nature. Many patents have been granted for plants; the reception area for Doll's group proudly displays a photo of the ten-thousandth plant patent being awarded in 1997 for a new type of geranium. Still, plant varieties were treated as special cases, and until recently other living organisms were not considered patentable.

The development of recombinant DNA technology in the 1970s challenged the patent system to extend its reach and grant applications for patents that covered descriptions of DNA sequences and organisms created through genetic engineering. Genes clearly are not novel or invented by man, but applications for DNA patents have been interpreted as requests for protection of new chemical compounds created by human endeavor rather than naturally occurring chemicals. Disclosure of a DNA sequence's molecular structure is considered sufficient to obtain a U.S. patent as long as the application includes some reasonable description of what the sequence might do in a living organism. In practice, even vague speculations about the function of an isolated portion of DNA have been adequate. The USPTO has been deluged with close to two hundred thousand applications covering any DNA strands that scientists can pull out from humans and characterize, a simple task with the gene-sequencing biotechnology currently available.

The patent office was not always generous in granting these biotechnology patents, but the largesse of Doll's team today has been facilitated by court cases decided before his tenure that established the patentability of genetically modified organisms. The most significant litigation in this area was Diamond v. Chakrabarty, a case initiated when Ananda Chakrabarty, a research scientist working for General Electric, tried to patent a bacterium that could break down crude oil following a spill. This bacterium had not previously existed in nature, but the application was rejected by the patent office because it considered bacteria to be complete living organisms and thus not patentable.

The case moved through the appeals courts and in 1980 on to the U.S. Supreme Court, which ruled by a narrow (five-to-four) majority that the intent of the intellectual property clause in the U.S. Constitution was to encourage rather than impede the granting of patents. The justices interpreted the key legislative clause "made by man" to include living things modified by human intervention, and supported the patentability of any organism made with recombinant DNA. The Supreme Court emphasis on "made by man" over "product of nature" was further broadened in 1988, when the patent office granted a patent for the so-called Harvard mouse, a bioengineered mouse with a heightened susceptibility for breast cancer that was developed as a tool for cancer research.

There are no longer significant conceptual legal cases pending about the broad patentability of genetically modified crops or bioengineered livestock, but there is more litigation between companies concerning the breadth of biotechnology patents. From 1991 to 1997, 48 cases were filed in which one company sued another over how patent rights for genetically modified crops were being exercised, a result of the patent office's having issued many overlapping patents. A typical lawsuit involved two companies, each holding its own patent on a different variety of corn engineered to express the same Bt protein. The arguments went like this: The plaintiff claims that its patent covers all varieties of Bt corn as well as all other Bt crops. The defendant argues that its variety of Bt corn in which the same Bt gene was inserted by a different process is not excluded by the defendant's patent, and is protected by its own patent. Further, litigants dispute whether other crops such as Bt cotton or potatoes are covered by patents for Bt corn.

John Barton from Stanford Law School is an expert on plant biotechnology patent law, and in a 1998 article he described the state of the legal playing field: "There are so many broad and fundamental patents that, in essence, every major actor may be violating a patent held by every other major actor." This situation is clearly unstable, but few if any of the lawsuits initiated by companies will proceed to trial. Rather, the purpose of these legal maneuverings is to inhibit competition by tying up companies in litigation, establishing broad patent protection for the last company left standing after the costly litigation process.

The vast majority of inter-corporate patent lawsuits have or will be settled in the traditional ways by which feuding businesses eventually make up. Some companies have divided up the biotechnology turf through complex licensing agreements. Others resolve their differences when larger corporations buy out smaller start-up companies through friendly offers that in the end are more appealing than a slow death through prolonged litigation.

Most commonly, biotechnology litigation is foreplay before merging. Patents were meant to provide protection by excluding competitors, and this philosophy is working, although not quite in the way the Patent Act intended. Rather, the elimination of competition is a side effect of mergers. After a merger, the resultant mega-company winds up with a broad spectrum of biotechnology patents, upheld not by litigation that determined the right of exclusion but by corporate consolidation that rendered competition moot.

The biotechnology industry is delighted at the relationship that has developed between corporations that apply for exclusive rights to genetically modified organisms and the patent office, which has been more than willing to grant that protection. Patent protection for GM crops has been an essential element in the development of the multibillion-dollar agricultural biotechnology industry, but not everyone is satisfied with this arrangement. There has been a significant undercurrent of protest that comes down to the simple question of who owns the earth's genetic resources, and it is an issue that has its roots in the earliest legal and political wrangling about licensing or patenting plant varieties.

Exclusive corporate rights to seeds were not always ensconced in American law. Instead, they evolved along with wholesale changes in U.S. agriculture during the late nineteenth and early twentieth centuries. Before that, farmers grew and saved their own seeds, devoting part of their plantings each year to producing seed for the subsequent year's crop. But then increasing mechanization and more efficient transportation changed American agriculture from subsistence farming to a market-driven, export-based industry. Farmers had decreasing resources to devote to producing seed for their next year's crop.

Simultaneously, scientists were developing more sophisticated methods of producing new crop varieties, working in a growing network of land grant colleges and U.S. Department of Agriculture laboratories with a mandate to improve farming. Cary Fowler, a specialist in international agriculture, described in his 1994 book *Unnatural Selection* why American farmers gradually allowed the experts to take over their traditional practice of growing and saving their own seed: "As skilled as the intelligent, observant farmer might be in plant selection and simple breeding, there was an incomplete understanding of the mechanisms of genetic inheritance. The advance of genetics created a wall between farmer and scientist which could not easily be scaled. Rapid progress would henceforth be made by scientists armed with genetics, mathematics, written breeding records and pedigrees, the use of statistics and a growing private and federally supported research establishment."

A mix of science and economics drove the transition from saved to purchased seed. New varieties emerging from government and eventually private laboratories showed improved yields and better resistance to pests and diseases than farmers' own selected seeds, and it became more profitable for a farmer to buy these seeds than to grow inferior seed on land that could more economically be devoted to crop production itself. The marketplace responded, and by 1918 60 percent of the seed used by farmers was purchased rather than home grown, produced by three hundred companies and sold through over thirty-five million catalogues mailed out every year to farms in the United States.

Most of the first plant varieties made commercially available to American farmers were developed in government and university laboratories, and for this reason were considered to be in the public domain and thus could be sold by any company. However, private companies quickly caught on to the advantages of inventing their own seed lines; any company that could produce a successful seed line would gain that important marketing edge of being the sole source for a unique commodity. But this private domain strategy depended on protection of intellectual property, and the increasing involvement of industry in seed production inevitably led to lobbying for some type of proprietary protection for the new varieties.

Industry's efforts to have plants protected by patents were unsuccessful initially. Congress continued to legislate as if the United States was an agrarian and small-business society, although the experiences of most Americans were becoming increasingly urbanized and corporate. Lawmakers were reluctant to permit patenting of food, fearing a political backlash from the heartland of America, which still considered farming to be the small-scale, rural way of life it had been rather than the multinational commercial endeavor it was becoming.

The nursery industry was the first to lobby successfully for patent protection, and those efforts led Congress to pass the Plant Patent Act of 1930, which allowed plant breeders to patent new varieties of fruits, berries, and roses produced from grafts or cuttings. But it was not until 1970 that Congress extended patent coverage to seeds by passing the Plant Variety Protection Act. Diamond v. Chakrabarty further cemented the right of industry to patent organisms, and today companies that produce genetically modified crops enjoy the full benefits of protection for their products.

For the companies, patent protection for the intellectual property represented by GM crops has become an essential component of their research and marketing strategies, while for critics it has become an unethical barrier separating farmers from their rightful access to seeds. The corporate perspective is a simple one: it can take hundreds of millions of dollars to make a genetic modification in a crop and bring it through the regulatory process to market, and this level of investment is justifiable only if a patent protects that investment from infringement by competitors.

GM crops are particularly susceptible to intellectual theft, because any competent plant breeder can cross a new GM crop containing the novel gene with his own variety. Further, a gene placed into one GM crop can be removed by another company's scientists and reinserted into the same or a different crop, and anybody who does this gets the benefit of many years or even decades of research in a very short time. Thus patent protection for genetically engineered crops is even more imperative than such protection for traditional varieties, because it is easier and quicker to steal a given genetic modification than to engage in the complex and long-term breeding schemes required to produce a conventional variety.

Patent protection for genetically engineered crops has not been of great concern to most farmers in developed countries, for whom annual seed purchases have been common practice for generations. Nevertheless, there has been vocal opposition from those who do not accept the basic premise of patenting living organisms and who are particularly concerned that GM crops exported to developing countries will destroy established farming practices and diminish the diversity and availability of traditional crop varieties. Their campaign against GM crops has failed to make any inroads in the legal system, either in the United States or overseas, so their focus has shifted to the court of public opinion. Their strategy is simple: lobby international trade and agriculture organizations to recognize the fundamental right of farmers to have direct access to seeds.

The campaign for farmers' rights is being conducted by a dizzying array of privately funded, nongovernment organizations, most of which include the weighty term "Institute" or "Foundation" in their titles to establish credibility. In the United States they include the Edmonds Institute, the Foundation on Economic Trends, the Institute for Agriculture and Trade Policy, the Rural Advancement Foundation International, the Institute for Sustainable Development, and innumerable others. Their members include lawyers, scientists, economists, and farmers disenchanted with conventional agriculture and agribusiness, and their mission is to force the agencies that negotiate and regulate international trade to decrease intellectual property protection for genetically modified crops.

Their strategy has been to produce position papers, hold forums, and engage in protests at trade meetings, rather than pursue legal action. Court challenges to intellectual biotechnology property may be feasible on a national basis, but global treaties make it difficult to pursue cross-border patent lawsuits. Thus these protesters' attention has turned to lobbying international groups like the World Trade Organization (WTO), the Food and Agriculture Organization (FAO), and the United Nations that are setting new global trade rules.

These organizations and national governments have created a tangled web of treaties and agreements with the objective of removing obstacles to international trade, and these efforts have included attempts to regulate intellectual property rights. Historically, patents or licenses for crops have been highly variable from country to country, with the strongest protection in North America and Europe and the weakest in the developing countries of Africa, Asia, and South America. Further, inconsistent patent regulations covering foreign inventions within each country have made intellectual property protection problematic at the international level.

The contemporary drive to bring trade under global jurisdiction has the objective of overcoming the previously idiosyncratic and patchwork rules under which nations made their individual trade agreements. New international treaties developed during the 1990s, such as the Convention on Biological Diversity, the International Undertaking on Plant Genetic Resources, the Trade-Related Aspects of Intellectual Property Rights (TRIPS) treaty, and the General Agreement on Tariffs and Trade (GATT), force countries to conform to international standards on intellectual property if they want to participate in global trade. Most nations have signed these treaties and by doing so have agreed to the broad principle of international patent coverage for crops, including genetically modified varieties, that are covered by treaty provisions.

Not everyone agrees that this policy is in the best interests of farmers or the public. The opposition viewpoint has been succinctly articulated by Kristin Dawkins, an analyst for the Minnesota-based Institute for Agriculture and Trade Policy, one of the organizations opposed to the growing corporate influence on agricultural trade. She wrote in a 1999 article for *In Focus* magazine: "Intellectual property rights grant inventors monopolies in exchange for their socially valuable innovations, a privilege that the United States interprets as a corporate right to privatize plants, animals, and other forms of life. Monopoly control of plants is contributing to the destruction of food security and public interest research, as well as to the loss of biological diversity and ecological health . . . Virtually all the world's nations have lost their right to determine the balance of private and public benefits designed to meet national goals. Instead, they must comply with a single international standard designed to open their markets to transnational corporate interests."

One objection to international intellectual property rights for plants is the potential for a cascade of genetically modified crops to overwhelm local varieties. This issue pits the use of diverse, locally adapted folk varieties selected by indigenous farmers over many millennia against the use of a few homogenous genetically engineered varieties sold by multinational biotechnology companies and advertised as providing increases in yield and protection from pests. The concern is that the corporate-selected seed will seduce farmers into abandoning their traditional varieties, thus reducing crop diversity and diminishing the available genetic pool for future generations of farmers to select from.

Another issue involves farming practices themselves, and the current contrast between farmers in developed countries, who purchase seed annually, and those in third world countries, who continue their traditional practice of saving seed for their next year's crop. Farmers who save seeds from genetically modified crops are the ultimate nightmare for biotechnology companies, because corporate profits depend on yearly seed purchases. Thus seed sale agreements for genetically engineered crops include provisions that require farmers not to save seeds for subsequent crops, creating an annual dependency on buying new seed that horrifies supporters of traditional, small-scale farming.

Seed saving as a political issue heated up in 1998, when the Delta and Pine Seed Company was granted a U.S. patent for the so-called Terminator gene. This gene functions by producing a toxin that kills seeds if they germinate. When included in a GM crop, the Terminator provides an ironclad guarantee that farmers cannot violate patent rights by replanting second-generation seeds. The Monsanto corporation immediately saw the benefits in moving beyond agreements into the more forceful and irrevocable protection provided by the Terminator technology, and quickly began negotiating to buy Delta and Pine to obtain the Terminator patent. The uproar about Terminator from the antibiotechnology movement became too intense even for Monsanto, however. The company soon withdrew the Terminator offer and publicly promised not to insert the gene in any of its products.

A third, and possibly the most fundamental, farmers' rights issue of interest to activists today is covered by the colorful term "biopiracy." Rights organizations are demanding that farmers be compensated for the development of modern crops by generations of their ancestors who selected and modified wild plants, only to have these crops coopted for genetic modification without payment. The raw plant material that serves as the substrate for genetic engineering was developed through ten thousand years of agricultural history, and the farmers' rights movement argues that these diverse varieties should be considered intellectual property and be accorded the same patent status as genetically modified crops engineered by biotechnology companies.

José Esquinas-Alcázar of the Food and Agriculture Organization office in Rome has referred to the intellectual property disparity between developed and developing countries as a conflict between "technology-rich" and "gene-rich" societies. Biotechnology corporations exploit the gene pool from traditional crops, defined by the FAO as those "arising from the past, present, and future contribution of farmers in conserving, improving, and making available plant genetic resources, particularly those in the centers of origin/diversity." Thus, farmers' rights advocates say, the inventors of genetically engineered crops should provide compensation to the breeders of conventional crops and their descendants.

The farmers' rights movement has a compelling but not easily winnable case. Biotechnology corporations have not completely dismissed this concept, since they do not want to alienate the potential markets for GM crops in developing countries. But they also have been wary about their own access to patent protection if they very strongly oppose the concept that traditionally selected plant varieties have some inherent intellectual property value. Thus trade agreements slowly have been moving toward compromise positions that support mechanisms like seed banks to conserve traditional crop varieties, and also propose granting some compensation to farm communities for the use of local plants in genetic engineering.

Perhaps the most ambitious international agreement concerning the sharing of profit and intellectual property between developed and developing countries has been formulated by the UN-sponsored Convention on Biodiversity. A document issued by the convention outlining broad principles for an agreement was signed by virtually every country on the globe except the United States at or shortly after the 1992 Rio Earth Summit, and the almost-final draft guidelines were released late in 2000. The stated intention of this document is to "provide a framework for the appropriate access to genetic resources and the fair and equitable sharing of the benefits arising from their utilization." Benefit sharing is broadly defined as "sharing intellectual property rights with the stakeholders that contributed to the conservation of these genetic resources or to the scientific research and development based on these genetic resources."

It is hoped that this high-minded agreement will be expressed in real terms through global efforts to conserve plant varieties in seed banks, programs to maintain indigenous agricultural practices, the development of workable ways for traditional farmers to save even genetically modified seed, mutual participation in product development, shared economic ventures, the implementation of provisions for affordable access to biotechnology for farmers in developing countries, and joint ownership of patents or license fees for commercialized products. The more radical fringe would like to go further than the convention, and develop a research system in which governments rather than corporations would fund crop development and new varieties would be freely accessible to all.

This scenario is highly unlikely to materialize, and even the vaguely worded concessions about sharing intellectual property protection and participating in joint ventures articulated by the convention may not translate into meaningful actions. In spite of the lofty and carefully negotiated rhetoric in the Draft Convention on Biodiversity, the key clause remains Article 4.1, added largely to get the United States to sign: "The Guidelines are voluntary." The language of the international community may be inclusive, but tangible mechanisms to implement farmers' rights remain elusive.

Issues concerning the patenting of genetically modified organisms are legally complex, economically significant, and fraught with implications for how we manage agriculture and trade. Discussions of these intellectual property matters also give rise to profound questions about the appropriateness of patenting living organisms and who should own the rights to life itself.

Critics of patents for new organisms tend to focus on the practical issues that arise from extending intellectual property rights to include genetically engineered organisms, but some also express ethical and moral concerns. Debra Harry is a Northern Paiute from Nevada and executive director of the Indigenous Peoples Council on Biocolonialism, an American group formed to protect the genetic resources developed by native peoples around the world from biotechnology. She was one of the speakers on a panel about patenting life at the "Big Money, Bad $cience" meeting in Vancouver, British Columbia.

Harry and others involved in the movement against patenting crops have focused on the rights of the indigenous peoples who developed agricultural resources and the integrity of all forms of life. During her lecture she highlighted the ethical dilemma that ensues when culture and nature collide with technology and intellectual property protection: "We see a tremendous clash of values and violation of collective rights of indigenous peoples . . . When it comes to ethics, they fly right out the door when scientists are collecting genes from around the world. For us genes have spirit. For scientists they're just cells. All life has its own integrity and shouldn't be subject to that kind of manipulation."

Morality issues have not gone unmentioned in the global trade arena, either. Indeed, the only exit clause included in the international TRIPS treaty concerning intellectual property rights involves morality. Patent protection can be overridden by a country if it damages human or plant health, induces serious environmental damage, or is morally objectionable. Nevertheless, this clause has yet to be invoked anywhere in the world, and certainly not in the business-friendly environment of the U.S. Patent and Trademark Office.

I asked John Doll whether an ethical issue had ever been raised successfully in challenging any American patent. His reply was characteristically direct: "In my twenty-six years in the patent office I have never seen a morality rejection made. I don't know of any case law that is directly on point to morality in biotechnology in the United States . . . We deal with statutes and we generally don't get involved in ethics. We almost always sidestep the issue, and a lot of people have tried to draw us into that discussion. Our point is that we are an administrative agency; statutes are passed by Congress, they are interpreted by the judicial system, and we administer the law."

The most recent attempt to force the USPTO to consider a morality issue involved the application for a patent for a human chimera submitted in 1997 by the cellular biologist Stuart Newman of the New York Medical College and the prickly social activist Jeremy Rifkin, head of the Washington-based Foundation on Economic Trends. A chimera is a creature that combines two or more animals, a term derived from a mythological Greek creature that had a lion's head, a goat's body, and a serpent's tail. Newman and Rifkin's application was for a process to merge human and animal embryos, transfer the chimerical cells to an animal or a human surrogate mother, and let the resulting creature mature and be born as a new life form.

Patent law does not require that the deed actually be accomplished; it requires only that applicants demonstrate its feasibility.

A human chimera is not as far-fetched as it seems. A few years ago scientists created a "geep," part sheep and part goat, and a chimp-human "humanzee" could be created with similar methods. However, the point of Newman and Rifkin's application was not to seek patent protection for a creature they actually were going to create but rather to force the USPTO to issue a morality rejection, which could then be used as the basis for further legal action to overcome the court decisions allowing patenting of living organisms that began with Diamond v. Chakrabarty in 1980.

Doll and the USPTO managed to sidestep the morality issue when rejecting the original human chimera application as well as two resubmissions. Rather than pronouncing a human chimera to be moral or immoral, Doll told me that "the office chose not to make a morality objection because the office had so many other good rejections . . . When we looked at prior art and enablement, we didn't feel it was necessary to deal in an area that is really unsettled in U.S. case law." Newman and Rifkin are persistent; in 2000 they sent in their third resubmission, attempting to counter all of the USPTO's reasons for rejection except that of what Newman politely calls "inappropriate subject matter."

The human chimera rejection was predictable from patent history, which favors the practical inventor over the moral objector. The U.S. Patent and Trademark Office has always kept its deliberations focused on utilitarian matters, refusing to consider the more controversial ethical and moral issues. Patents have been denied, but only because the processes they describe have been carried out before (prior art), the inventions are not described well (lack of enablement), or they cannot be demonstrated to be feasible (lack of credibility).

Thus although patents can legally be denied on moral grounds, none has yet been refused for that reason. It seems unlikely that ethical objections to genetically modified crop patents will be successful if applications for human chimeras cannot force the

patent office into the ethical realm. If a "humanzee" won't generate a moral rejection, Bt corn or herbicide-resistant canola certainly won't, either.

Patents have one fundamental purpose, to take the genius
of invention and provide exclusive rights to its fruits in order to
fuel economic growth. Patent law has evolved with this benefit to
society in mind, and favors the inventor over the intellectual tres-
passer and the exclusion of competition over morality considerations. The offices around the globe that issue patents do so under
the direction of and the laws passed by their governments, and so
far the nations of the world have agreed to allow the intellectual
property represented by genetically engineered organisms to be
protected under both national patent laws and international treaty
agreements.

Policies protecting inventors throughout the world have been
taken the furthest by the U.S. Patent and Trademark Office and
American trade negotiators; both are unabashedly pro-business
and are proud to grant or promote bioengineering patents. And
both have carefully avoided being drawn into morality arguments. The intellectual property issues they have considered have
dealt not with whether biotechnology is right or wrong but only
with how best to protect genetically modified crops so as to generate wealth for their inventors, for corporate producers, and for
shareholders.

Key bureaucratic policies, legal decisions, and trade agreements
have protected crops and other genetically engineered organisms
from competition and from claims by indigenous peoples to some
share of the proceeds from that intellectual property. Proprietary
claims on living organisms have been disputed through the
courts, but those challenges have failed to limit patent protection
for bioengineered organisms. Critics have lobbied hard to pro-

mote universal and patent-free access to the world's genetic resources, but have not influenced nations to take tangible steps
that provide open access to bioengineering technology or income for indigenous peoples from bioengineered products. Critics opposed to genetic engineering have played the moral card,
but no court or international body has issued a ruling rejecting
bioengineered products on ethical grounds.

> Today's overwhelming volume and variety of information makes it possible—by selecting and connecting data points carefully—to paint practically any picture of the world and make it seem accurate. So the pictures we paint are often more a reflection of our deepest personal orientation, especially of our basic optimism or pessimism, than of empirical evidence.
>
> **Thomas Homer-Dixon,** *The Ingenuity Gap,* **2000**

There'll Always Be an England

Traveling between North America and Britain these days can be a perplexing experience for scientists, especially ecologically oriented biologists like me who also have strong affinities to agriculture. The world of North American farming today is bullish on biotechnology, with only occasional islands of environmental concern surfacing above a vast sea of crops that increasingly are genetically modified. The United Kingdom is in an alternate universe, a world in which the fundamental balance between scientific advances and ecological ethics has evolved in a very different direction.

Intense protest against genetically modified crops has become as prominent a fixture of the British landscape as thatched cottages, country pubs, and fish and chips. Menus and window signs in virtually every pub, restaurant, bed and breakfast, and takeaway store proclaim that no GM food is served here, and not a day passes without a significant story on the GM controversy in the hyperactive British press. The campaigns for and against GM have taken on their own bombastic momentum, and even the relatively staid BBC news reports on the issue are difficult to distinguish from *East Enders* and *Coronation Street,* the soap operas that rivet most Brits to their tellies in the evening.

The impact of the anti-GM forces has been decisive. Genetically modified crops are going nowhere in Britain, whereas GM crops have become predominant in many sectors of American and Canadian agriculture. The future of North American farming increasingly and almost inevitably seems to depend heavily on these new techniques, and protest has had relatively limited success in disrupting their use. Britain is evolving in a different direction, one in which public pressure has restricted biotechnology in farming to an inconsequential role in British food production.

The intensity of the opposition to GM crops in Britain has taken farmers, the biotechnology industry, and the British government by surprise. Scientists on both sides of the ocean also are taken aback by the strength of the opposition, and puzzled by the protesters' lack of engagement with scientific issues. The opposition to GM crops in the United Kingdom has revolved around broader issues rooted in British politics, history, ethics, and views about nature.

Whatever the roots, the pro- and anti-GM cultures remain isolated from each other. Sue Mayer from GeneWatch, a British group active in GM issues, put the situation well: "I inhabit two universes in the U.K. There was one universe which I was inhabiting yesterday; last night for example was a public meeting on GM issues, deep anger and passions. It's quite extraordinary because there's a culture of public meetings at the moment; you get hundreds of people in all sorts of weather. Then you have the regulators and the biotech industry in a different universe, and they simply do not understand or want to understand or have been able to grasp the challenges that are coming from the public."

The GM controversy in Britain comes down to a series of intense confrontations between big business and government on the one hand and a shrill but effective environmental lobby on the other. The arena in which the GM battles have been fought is that of public opinion, and the weapons of choice have been cleverly worded and carefully produced reports from both sides extolling or condemning GM crops. The environmentalists have relied on emotional and politically based arguments against the use of biotechnology in agriculture, while industry and government have relied on scientifically based reasoning and projections of the economic advantages of using GM crops. So far, politics has won out over science, and GM crops and foods in Britain have been effectively routed.

The anti-GM movement has been led by three environmentally correct organizations, although there also are legions of smaller groups that have participated in the protest. Greenpeace, Friends of the Earth, and GeneWatch have been the principal opposition forces, with Greenpeace and Friends of the Earth being the most radical and confrontational. GeneWatch espouses views similar to those of the other two groups, but uses less inflammatory rhetoric.

On the pro-biotechnology side are the expected promoters of GM, including agrochemical and seed companies and British government offices with responsibilities for science, agriculture, economic growth, and trade. These advocates of GM crops speak out freely, but also rely on the activities of independent groups not directly connected with them. Prominent groups that support agricultural biotechnology include the Royal Society, the Food and Drink Federation, and the Nuffield Council on

Bioethics. These organizations have issued reports, some more enthusiastic than others, that express support for agricultural biotechnology in a tone that is generally less raucous and more neutral than that employed by government and industry groups.

The most clear-cut issue separating the pro and anti sides has been the problem of benefit versus risk, with the pro-GM advocates arguing that GM crops offer considerable benefits with no real evidence of damage to date, although they cover their flanks by supporting continued research and monitoring. The opposition raises the specter of largely unspecified negative effects that have yet to ensue, and argues that the benefits are vastly outweighed by the risks and that the goals of genetically modified agriculture can more easily be achieved through changes in political and economic policies than through the use of insufficiently tested science and technology.

But although both sides have focused on perceptions of benefits and risks, at the core of their dispute is a fundamental disagreement about the role of industry and science in shaping the world. The pro-GM advocates rely heavily on science to make their points, producing glossy reports that rely on statistics-laden tables, flow charts, long lists of benefits, and a lack of proven side effects to make their case. Theirs is a sophisticated and carefully crafted public relations campaign, sprinkled with socially friendly terms such as "food future" and "the way forward," and phrases coopted from the environmental movement such as "stewardship" and "biodiversity."

The benefits of agricultural biotechnology touted by GM proponents are repeated so often that they have become like mantras. After reading report after report, I found it easy to imagine the suits and the lab coats meeting together in the halls of industry and the offices of government three times a day in a quasi-religious ritual, chanting something like "These are the benefits of our work: fewer pesticides used, increased yields, drought resistance and flood tolerance, increased vitamin A in rice, functional foods with nutritional benefits, cleaning up environmental contaminants. These are the benefits of our work . . ." These indeed are real benefits, but in spite of the considerable advantages of GM crops and the lack of clearly proven risks to date, the British public has not been converted.

Rather, opposition to GM crops has become a national obsession because of the more emotional and much more successful campaigns conducted by the protesters. They, too, issue weighty reports, but it is their Web sites, press releases, guerilla tactics, and media coverage that have captured the hearts of the British people. The opponents' language contrasts starkly with the proponents'. Phrases such as "a grubby experiment with our health," "genetic pollution," "biopiracy," "imposed without consent," and "forcing this new technology down our throats" pepper the anti-GM broadsides. Further, the opposition has capitalized on the public's love of theater. On street corners, the opposition has staged attacks by flamboyant protesters wearing white environmental protection suits on trial plots of GM crops that resonate with the entertainment-loving British public more than the ponderous press conferences held by GM proponents.

The success of the GM opposition has been dramatic, and the depth and breadth of the anti-GM movement have been difficult for the more technology-friendly North American public to comprehend. In Britain today, there are no commercially planted GM varieties, and most of the few field trials that have been authorized by the government have been disrupted by protests. As for food itself, the occasional imported foods that might contain GM products are so labeled, and they have proven increasingly unpopular with consumers. By contrast, in North America the majority of canola, soybeans, and cotton and a third of the corn planted today are genetically modified, and that trend is continuing with other crops. In addition, most North American processed foods contain some GM components, but there are no warning or informational labels on food packages. Food in restaurants also originates from biomodified sources, but it is rare to find a restaurant door or menu displaying an anti-GM warning.

Sue Mayer was originally trained as a veterinarian and pharmacologist, but her early work with seal rehabilitation and her growing concerns about environmental issues led to a stint with Greenpeace as its director of science. She moved to Tideswell in the Peaks District of England for personal reasons in 1994 and founded her own organization, GeneWatch. Funded through donations and contracts with industry, government, and other nongovernment organizations, GeneWatch was established to evaluate the ethics and risks of genetic engineering and to question how, why, and whether the use of genetic technologies should proceed.

The GeneWatch office reflects the organization's dual emphasis on scientific analysis and politics, with high-tech publications like *Nature Biotechnology* displayed side by side with anti-apartheid posters and new-age environmental books. When I met with Mayer, she began our conversation about biotechnology by talking about the multinationals: "People are hacked off with large multinational companies, and don't think it's being done in their interests, it's in the interests of the large corporations. There is this kind of feeling, you know, how far will we go to make money, where will it end, what are the boundaries on that? The feeling that it's driven by commercial interests I think makes it feel very dangerous and deeply unethical to people."

Some of this distrust of corporations can be attributed to the strong left-wing political tradition in the United Kingdom and Europe that is deeply suspicious of business. Even so, "multinational" has become an unusually pejorative term used specifically in the GM debate to disparage the corporate agricultural biotechnology community. Like many such terms bandied about by lobby groups, "multinational" is perceived as a shorthand expression, in this case a code word for "American chemical companies," particularly Monsanto. However, the reality of multinational corporate influence is considerably more complex.

The few GM crop trials being conducted in England today are indeed using seed developed by multinational companies, but their home offices, corporate connections, and economic ties go well beyond the United States. The three major crops being field-tested in Britain during 2000 in a very limited way were oilseed rape, forage maize, and sugar beets. The genetically modified rape and maize seeds were bioengineered by Aventis, a primarily French company formed by the merger of Hoechst of Germany and the French Rhône-Poulenc, and the sugar beet seed development originated as a joint venture between Monsanto of the United States and Novartis (now Syngenta) of Switzerland.

Many other international agrochemical companies have been involved in developing GM crops, but the fifth-largest seed company in the world, AstraZeneca, has some British roots; it was created in spring 1999 when the partly British company Zeneca Seeds merged with the Swedish company Astra Pharmaceuticals. The company was burned badly in its previous life as Zeneca after it introduced the first GM food into Britain, the Flavr Savr tomato. Canned paste from these tomatoes was the first target of British anti-GM protest, which drove the product from British shelves in spite of the fact that it was 20 percent cheaper than paste made from conventional tomatoes.

The former Zeneca Seeds no longer focuses on British markets; instead it has turned its corporate attention to GM crops for developing countries, where the company hopes to encounter less restrictive regulations and less protest than in Britain. The company's British connection was submerged even further with the November 2000 merger of Novartis and AstraZeneca into Syngenta. It is difficult to pinpoint why the British public is unusually suspicious about this contemporary trend toward international biotechnology giants, but perhaps the loss of any obvious British identity through mergers has amplified the lack of respect with which the public regards multinational companies with genetically modified products.

Some of this aversion to multinational companies may be attributed to the British realization that multinational no longer means the British Empire. The decline of British economic influence during the twentieth century was dramatic, and there remains a lingering sense of loss in the United Kingdom that affects the British view of new technologies coming from overseas. From 1900 to 1995, the British share of world trade dropped from 33 percent to less than 5 percent, and, perhaps more significant, Britain has become a net importer of scientific and technological advances. Biotechnology is a prime example. The structure of the genetic material itself, DNA, was jointly discovered in 1953 by an American and a British scientist, James Watson and Francis Crick, but the economic development of DNA-based technologies took place offshore, deeply rooted in academic science and corporate research laboratories from the United States and continental Europe. Even Zeneca Seeds entered the GM market by purchasing companies from overseas to acquire their technology.

Whatever its origins, the British antipathy toward multinational companies runs deep, and the British have similar attitudes about international quasi-government groups such as the World Trade Organization. The core of their distrust is the belief that the WTO is manipulating independent-minded governments into ceding authority over decisions about importing food and growing GM crops to a centralized body that they view as favoring American business interests. Margaret Llewellyn from the Biotechnological Law and Ethics Institute at the University of

Sheffield summarized the resistance to WTO policies succinctly: "The view in Europe is that the World Trade Organization is run by American accountants."

The British distrust of corporate biotechnology, globalization, and international trade organizations is also linked to British attitudes about developing countries. Increasingly intense battles are being fought over the development of functional foods designed to improve nutrition and health in the third world. Companies are racing to produce these functional GM crops to expand their product lines and win over public opinion, but opponents view the purveyors of these nutraceutical and pharmaceutical products as being even more perverse and economically manipulative than the purveyors of the present generation of pest-resistant crops.

Functional foods contain ingredients that provide health benefits beyond that of normal nutrition, and advocates point to their win-win potential to improve conditions in developing countries while earning money for investors. To enhance foods and nutrition, GM techniques are being used to produce crops with increased vitamin, mineral, and protein profiles, modified fats and oils with healthier components and reduced processing requirements, increased starches and sugars, and crops that are free of allergens that can be life-threatening. In the realm of medicine, the prospect of inserting genes that code for vaccines into food has promise as a way of delivering high-technology medical products to low-technology countries.

But even so, these functional foods are strongly opposed by protesters. The anti-GM British lobbyists are not, of course, hard-hearted or insensitive to the plight of dying third world children, but they do question whether functional foods really confer the touted health benefits or are just another mechanism for exploiting the developing world in the quest for corporate profit.

The cynicism with which functional foods are viewed in Britain was apparent at an April 1999 press conference held to publicize a report critical of these upcoming GM products, titled *Biotech—The Next Generation: Good for Whose Health?* The site of the event was one of the most hallowed in London, Westminster Hall, the oldest part of Parliament, from which England has been governed since 1099. Plaques embedded in the stone floor commemorate famous impeachment trials, sentencings, and coronation banquets that have taken place there, and from on high gargoyles lining the upper walls gaze grimly down on the events below, as they have for almost a millennium.

The report was researched and written by Sue Mayer of GeneWatch and Sue Dibb of The Food Commission, an independent group that views itself as a consumer watchdog on food issues. Even a cursory reading quickly reveals their deep ambivalence about supporting science-based food and health initiatives for developing countries if they involve GM crops and corporate profits. The sound bite–friendly summary begins and ends with blunt anti-GM statements: "In a desperate effort to reverse its failing fortunes, the biotechnology industry and its supporters are putting their faith in the 'second generation' of genetically modified crops . . . There is evidence that GM crops with enhanced nutrition are being used to gain better public acceptance of the biotech industry as a whole. Most second generation GM crops which have been nutritionally altered will not offer direct benefits to consumers. The primary beneficiaries are likely to be the biotechnology companies developing the crops, food processors and producers, the animal feed industry and non-food industrial users."

Also speaking at the press conference were Joan Ruddock, a

Member of Parliament, Tim Lang, an agricultural economist, and Clare Joy, a representative of the World Development Movement, each of whom echoed these disparaging views. They talked about the exploitation of developing countries by desperate corporate monoliths. Tim Lang, from Thames Valley University, was particularly forthright: "What is driving them is that they are running out of markets, they are running out of the capacity to buy each other out, they are running out of the capacity to increase added value in your foods. What functional foods do is create an entirely new sector which is nirvana in economic terms. This is not a concern about saturated fats, but a concern about saturated markets."

The alternatives proposed by the opponents of functional GM crops stem from the belief that political, economic, and cultural change would be sufficient to solve hunger, poverty, and health problems in the third world. Not surprisingly, they trot out the oft-used and probably correct argument that enough food is produced around the globe to feed the current human population, so that if food and wealth were redistributed there would be sufficient grains, fruits, and vegetables to provide adequate nutrition for everyone in the world. They view the science and practice of biotechnology as a smokescreen obscuring the true agenda of the developed world, which is exploitation of the poor for corporate profit.

What appears unusual about this belief from the other side of the Atlantic is that it is held not only by the expected leftist fringe, but by the British public at large. Perhaps this British sensitivity to third world issues should not be surprising, given that the recently defunct British Empire was maintained for centuries by exploitation of the poor and undeveloped by the rich and technologically advanced. The current widespread reluctance in Britain to combat third world poverty and misery with exported biotechnological solutions appears to be influenced by the realization that Britain once was a colonial power that maintained its dominance through the use of superior technology and military force.

British attitudes about multinational companies and developing countries are not the only attitudes that have contributed to the formation of the majority anti-GM view in the United Kingdom. Another potent force in building the opposition to genetically modified crops has been a deeply ingrained distrust of government regulators.

The source of this distrust is not hard to identify. It lies in recent U.K. regulatory history. As Sue Mayer pointed out, "We've had a series of food safety problems in the United Kingdom which have been poorly dealt with by our institutions. Not only has there been a kind of elevation in anxiety over food safety, there's also been a complete fracturing of trust in the institutions that are supposed to protect people from [unsafe foods]. The institutional responses to GM crops are very similar to the responses to other kinds of food safety issues. There's an underlying distrust on the part of the public. There is also anger at the lack of consultation about these issues."

Mayer was referring to a number of incidents of food being contaminated by bacteria such as salmonella and *E. coli* and, more significantly, to the most horrendous threat to food safety, bovine spongiform encephalopathy (BSE). More commonly known as mad cow disease, it is one of a class of diseases that leave a spongy appearance in the brain which is visible when tissue is examined under the microscope. The causative organism is a subject of considerable debate, but it appears to be even more primitive than viruses, and is thought to be transmissible not only between cattle but also to humans who eat tainted beef. The human version is called Creutzfeldt-Jakob disease. It has killed about a hundred people since 1985, and many more are thought to be infected but not yet showing any symptoms. Besides its devastating effect on human life and health, the disease has inflicted overwhelming economic damage on the British beef industry: enormous numbers of cows have been slaughtered to prevent the spread of BSE, and many markets, both local and overseas, have been lost.

The British public and most experts believe that sloppy agricultural practice—adding ground-up cattle carcasses to feed— spread this disease in cattle, and that government food regulators should have prevented this from happening. Superficially, this issue has nothing to do with genetically modified crops, since not even the most rabid anti-GM opponent believes that GM crops carry disease-inducing organisms.

Nevertheless, the British public perceives the BSE debacle as yet another failure on the part of the regulators. It has so increased their lack of trust in their own regulatory system that even GM crop producers and regulators have become suspect. Government assurances that food containing GM crops is safe and that the crops themselves pose no environmental threat sound similar to earlier statements designed to reassure the British public that BSE really was not a problem, and to obscure its likely origins in sloppily prepared cattle feed. In addition, the plethora of committees with long names and confusing acronyms that have sprung up to deal with the GM issue are not reassuring, but alarming. Government or quasi-government organizations with such tongue-twisting designations as the Supply Chain Initiative on Modified Agricultural Crops (SCIMAC) or the International Service for the Acquisition of Agri-biotech Applications (ISAAA) are viewed as not much different from the Spongiform Encephalopathy Advisory Committee (SEAC).

There is a more specific issue that does evoke eerie echoes of BSE, and that is the use of GM crops for cattle forage, with the potential for an undesirable protein to get into the meat. There is no evidence to date that this has occurred, or even that there are undesirable proteins in commercially available GM crops, but nevertheless this possibility has caused a mini-panic. Jayn Harding from the British supermarket chain Sainsburys described reactions of consumers to this and other food safety issues: "They're saying they've lost confidence in the food industry, the government, everybody, and they can see something similar to the BSE with GM's. They're looking at it like 'You're playing around with my life and you don't know for sure that you can play around with my life safely, and until you can, I don't want you to.'"

Many British arguments about the implementation of GM science stem not only from concerns about safety and exploitation, but also from moral objections. To North Americans, this emphasis on ethics may be the most surprising aspect of British attitudes toward crop biotechnology, since the words "moral" and "ethical" rarely surface in North American debates about GM crops. While some North Americans are concerned about the ethics of using biotechnological applications in medicine or in human cloning, and some worry about the environmental side effects of genetically modified crops, few object to GM crops on ethical grounds.

I spent a morning at the Biotechnological Law and Ethics Institute at the University of Sheffield with Margaret Llewellyn, a lawyer and faculty member who specializes in biotechnology patents and intellectual property issues. Every surface in her office is piled high with files, and her shelves groan with an array of ponderous intellectual property books, such as *Practical Intellectual Property*, *Intellectual Property and the Law*, and *Principles of Intellectual Property*. Even her key chain carries out the theme; it holds a replica of the world's most famous sheep, Dolly, cloned and still living a few hundred kilometers north of Sheffield.

She described for me the history and role of moral considerations in science-based issues: "Morality is very much something that's part of European culture. Within politics, within literature, every element of the culture, morality and philosophy in the legal sense of jurisprudence, there's been a very strong moral theme and a very strong integration between science, education, and religion, way back to the Greeks and Romans. There's always been a thread of morality which has attached itself to every discipline, and so even if a new discipline arises somebody, somewhere, will ask what are the moral implications of this and will devise their own moral perspective . . . It's very clear at every level that morality is a requisite element for consideration and it's stated as being such. And I don't know whether or not that makes us a more moral society or makes us more of an intolerant society or if it makes us more of an arrogant and smug society because we claim to take morality into consideration."

Antibiotechnology arguments based on moral grounds endorse some practices and condemn others on the basis of their perceived value to humanity. Opponents of GM may admit that GM crops confer economic benefits on farmers and big business, and can have important uses in medicine, but in other ways they do not view these crops as beneficial for humanity. Llewellyn said, "There seems to be less of a problem with genetically engineering a pig to have a heart or a liver which is used for transplantation in human beings than there is about crops . . . You haven't got people going up to the Roslin Institute and setting Dolly free the same way you have organizations such as Greenpeace tearing up GM crops in fields where they were on trial. The explanation I see for this is that the medicines might benefit us, and medical research should be encouraged not only for selfish reasons but for nonselfish ones, because for somebody somewhere suffering might be alleviated. With the environment, the benefits are less personal."

Sue Mayer elaborated on this sort of seemingly capricious opposition to biotechnology: "It's not against all scientific endeavor at all, it's against the application and the perceived usefulness of it . . . People are extremely discriminating about applications of genetic technologies, so they tend to be very positive about applications in medicine, but science as applied to GM crops is seen as not really giving any benefit. They're savvy about science and its contingent nature."

This perspective has led to public acceptance and approval of many medical advances and policies, but continued resistance to the use of genetically modified crops. The first test-tube baby was created and born in Britain; Britain was the first nation to permit stem cell research to proceed with surplus human embryos created through in vitro fertilization; and both British and continental patent offices have granted patents for human cloning, which would certainly be refused by their counterparts in North America. A 1998 biotechnology directive from the European Union patent office made it clear that patents should not be granted if they sanction immoral acts, but human cloning patents have been permitted because of the potential of cloning to reduce suffering.

This is a highly utilitarian version of morality, where moral means beneficial to humanity, which, as Llewellyn put it, "is very different from a morality based on divine order, where God wishes us to issue a patent over this." Fortunately for multinational companies, plant varieties in Britain and the European Union are protected by license agreements rather than patents, whether they are genetically modified or conventional, so corporations have been able to sidestep patent issues about crop morality.

In the end, the most potent anti–GM crops force in Britain may be not any concern about food safety, exploitation, or even morality, but rather the British view of nature and of the impact

GM crops might have on the country's landscape. If there is any trait that unites the residents of Britain, it is a passion for rural settings and a close kinship with the countryside. The possibility that a gene from a genetically modified crop plant might jump into a wild weed and in some way change the natural order is viewed with horror.

Rustic Britain is indeed beautiful and accessible, and the passionate desire to protect it is both commendable and easily understood. Jeremy Paxman, a popular English interviewer and journalist, eloquently described the allure of rural Britain in his book *The English:* "It is the charm of small things; there is scarcely a geographical feature in the land that has any claim on world records. It is a place of tended beauties; the country lane, the cottage small, the field of grain, belong to a landscape that has been shaped by generations of labour. Its appeal is charted in fields and acres."

Beautiful and distinctive it may be, but it is arguably the most managed land on earth. The heartland of Britain is an extended country estate, with hardly a corner that is not farmland or pasture, and what remains is carefully tended to enhance its visual impact. The Britain of today may contain many wild plants and animals, but they have been rearranged and reorganized into an exceptionally human vision of nature.

It is not concerns about unmanaged nature that are driving the British opposition to GM crops, but rather a deep fear that GM crops could alter this familiar landscape. This passionate determination to maintain an artificially constructed nature is ironic, since even the severest predicted impact of GM crops on the environment would pale beside the changes that human management has already wrought in rural Britain. Yet the risks of environmental change from GM crops have stirred the normally staid British public into an unprecedented frenzy that has been a major force in halting the implementation of this new form of agricultural management.

The perceived environmental risks focus on the possibility that a genetically modified crop will jump from farmers' fields and go feral, or that pollen from a GM crop will pollinate a closely related weed species and genetically intrude into the unfarmed countryside. These are legitimate concerns, especially in Britain, where crops such as oilseed rape do have close unmanaged relatives that could be modified through pollen from GM crops. If so, such hybrids or the genetically modified crops themselves could displace existing flora as well as plant-feeding insects and the bird species that feed on them. What complicates this issue is that the long-term impact and magnitude of such events is impossible to predict. Such changes could just as easily increase biodiversity as decrease it, and could as well be beneficial as harmful.

But the mere likelihood of any sort of change has been enough to arouse the environmentally sensitive British public. Even Sue Mayer, who usually speaks in measured tones, became passionate when asked why she was devoting so much of her life to opposing GM crops: "If you were going to have some effect that would cause the loss of particular species, you can't help but think 'What are we doing? Why are we doing this? Do we have to do this? Could we not tread more lightly?' So there's also that kind of deeper questioning, as well as the rather pragmatic, practical stuff . . . There is something I think is more intangible about it, which is this whole issue of changing ecosystems in ways which you know are irreversible, unpredictable. I'm not being dippy about it, but these are important things which are incredibly difficult to articulate about how you feel about nature and environment."

The British bias against and current ban on genetically modified crops raises the question of which side of the Atlantic has it right on this issue, and whether North America will see a rising and eventually successful movement opposing crop biotechnology. As we have seen, the British opposition to GM crops includes components of suspicion about the motives of multinational companies, a long and complicated history with third world countries and development issues, a deep distrust of agricultural regulators, an unusual morality-based perspective on tampering with crops, and a passionate attachment to the traditional British countryside. These elements of opposition are not as intense in North America, where corporations are not as widely distrusted, developing countries are not primarily regarded as objects of corporate exploitation, the government systems that regulate food are more trusted because they have not been involved in as many food safety incidents, ethical issues about crops don't arise, and there is a less passionate desire to leave managed areas of the country unaltered.

When I am on one side or the other of the Atlantic, I find it easy to adopt either of these value systems. In Britain, I ponder the wisdom of moving too quickly to implement GM crop technology, but while traveling through the American and Canadian breadbasket states and provinces I wonder what all the fuss is about. Oscillating between the British and North American views made me realize how little either perspective is grounded in the science itself.

Scientists often dwell in a fantasy kingdom where the data alone will show the way, but the reality is very different. Ultimately, it is value systems and economic interests that determine whether nuclear power is embraced or repelled, antibiotics are overused or underutilized, genetically modified crops are planted in the ground or torn from the fields. Different perceptions of risks and benefits will always generate divergent perspectives, none of which can be considered definitively right or wrong.

As regards GM crops, North American and European societies have adopted dramatically different viewpoints, and it appears unlikely that either will change its view substantially. We face a long period of difficult negotiations and compromises about GM crops and biotechnology as agricultural trade issues collide with diverse societal and political values, not only in Britain, continental Europe, and North America but around the globe. Science may play a role in these discussions, and benefits may be balanced with risks, but the final outcome will depend more on cultural and societal beliefs than on the interplay of fact and data.

For the Good of Mankind

Protest needs symbols, and Monsanto has been the poster-villain for the antibiotechnology movement. It was one of the first companies to commercialize GM crops, and it entered the marketplace aggressively, heedless of the possibility that these new products might arouse strenuous opposition. Monsanto opposed food labeling vigorously, vehemently denied that bioengineered crops might have negative health and environmental impacts, required that farmers pay royalties and sign unprecedented agreements to plant their GM crops, used the draconian Terminator technology to prevent farmers from saving GM seeds, and issued a stream of press releases questioning the credibility of and demeaning the positions of GM opponents.

At first, Monsanto persisted in this unabashed advocacy for all things genetically modified. Its corporate strategy was to hunker down and weather the storm without attempting to placate the increasingly vocal and effective opposition. Monsanto came to symbolize the worst of the multinational corporations, with its reputation as a company that used brutal tactics, had an excessive focus on profits rather than ethics, and consistently behaved arrogantly.

Monsanto clearly underestimated the depth of protest and the opposition's ability to mobilize public opinion, but the company eventually joined the rest of the biotechnology industry in projecting a kinder, gentler image. Part of this recent public-friendly profile has been a conscious movement away from manufacturing GM crops designed to increase agricultural productivity, such as herbicide-tolerant and pest-resistant plants, and toward producing products with more socially relevant benefits. Consumers did not relate to or empathize with the first generation of farmer-focused crops; they viewed them as little more than glorified pesticides, and saw little benefit to and some risk in eating food produced from such bioengineered plants.

The second generation of genetically modified crops offers a more palatable and ethically admirable suite of products poised for commercial development. Monsanto and the few other large multinational companies that have persisted through mergers and anti-GM protests are preparing to market crops with improved nutritional attributes. They hope that the public will view these new crops as beneficial and not oppose their use.

Cynics say that the marketing of this new generation of crops is just more of the same, merely a more sophisticated way of inducing the public to accept exploitive technologies that are dangerous to human health and the environment and that generate excessive corporate profits. The biotechnology industry, naturally, adamantly disagrees. For Monsanto and the other companies involved in creating and marketing nutrition-enhanced GM crops, these bioengineered plants have enormous potential to feed the poor, improve human health, and provide direct benefits for developing countries.

Cherian George is Monsanto's technical director in charge of improved oilseed crops. He views Monsanto's new products as humanitarian aids: "Two to three hundred million poor people could clearly benefit. It's our commitment for the good of mankind."

■■ Catastrophic events in developing countries can be so dramatic that we overlook the daily and mundane. Visually overwhelming images of floods, earthquakes, wars, and famine routinely make the evening news, but while tragic they distract us from the widespread and persistent malnutrition that is the greatest challenge to daily life in the third world. If biotechnology can combat this pervasive threat by improving the daily food supply, it will make an important contribution to the betterment of the human condition.

Persistent nutritional deficiencies wreak untold havoc. Vitamin A deficiency is a good example: it causes blindness in children; exacerbates diarrhea, respiratory diseases, and measles; and often is fatal for children in developing countries. The statistics are dramatic: 134 million children throughout the world experience some level of vitamin A deficiency, 5 million experience permanent eye damage, and about 500,000 become permanently blind every year. Two million children under the age of five die each year from diseases associated with insufficient vitamin A in their diets.

Vitamin A deficiency is particularly serious in Asia, where rice is the primary cereal crop. Rice is a marvelously nutritious food, but it contains no vitamin A or the plant pigment beta carotene that we metabolically convert into vitamin A. Other sources of vitamin A such as milk, butter, and cheese are lacking in most third world diets, and the fresh vegetables that could provide vitamin A's beta carotene precursor are not affordable for many. Efforts to fortify rice and other foods with vitamin A or beta carotene by using traditional breeding methods or adding food supplements have not been successful enough to solve the problem, partly because of technical difficulties but also because of their high cost. Biotechnology has solved the problem by creating Golden Rice and bioengineered canolas and mustards that can deliver appropriate levels of beta carotene cost-effectively through genetically modified varieties.

Golden Rice had its origins in a meeting organized by the Rockefeller Foundation that led to an international research effort to genetically modify rice to yield beta carotene. The foundation initiated the International Rice Biotechnology Program in 1984, and since then has invested over $110 million in rice biotechnology research. The foundation invited a number of scientists to New York in 1993 for a brainstorming session, and among those attending were Ingo Potrykus of the Swiss Federal Institute of Technology and Peter Beyer from the University of Freiburg in Germany.

Beyer had previously discovered how beta carotene was produced in daffodils, and his laboratory had isolated three genes essential to its production. Potrykus's laboratory specialized in transferring genes into rice, and seven years after the two scientists met, they successfully inserted the daffodil genes that produce beta carotene into the East Asian *japonica* variety of rice. Beta carotene is yellow, and the enhanced rice is a golden hue when hulled, so it was called Golden Rice.

Canola and closely related mustard plants also have been modified to produce beta carotene. Unlike rice, these oil-producing crops naturally contain low levels of beta carotene, and researchers in Monsanto laboratories have engineered canola and mustard plants that overexpress their normally low beta carotene levels. Both the Golden Rice and the enhanced oilseed crops produce sufficient beta carotene to alleviate vitamin A deficiencies when the foods derived from them are consumed in quantities typical of those consumed daily by people in developing countries. In addition, beta carotene can be extracted from both sources and then sold inexpensively as a food supplement.

The science is no longer a barrier; bioengineered crops now are available that can dramatically reduce health problems caused by vitamin A deficiency. The agenda has shifted from engineering the plants to social, political, economic and intellectual property issues, and these have been considerably more complex to resolve than the science.

Golden Rice was difficult enough to invent, but the intellectual property issues surrounding its distribution appeared insurmountable. Potrykus and Beyer had signed sixteen material transfer agreements and licenses with ten different companies to obtain permission to use certain rice varieties, cloning vectors, and DNA transformation and amplification machinery in their research. In addition, the proprietary materials and methods they employed were protected by seventy patents held by thirty-one different companies in over thirty countries around the world. These agreements covered only the research, and did not permit the commercializing of any of the resulting products.

Potrykus and Beyer had always intended to share their technology free of charge by allowing public-sector breeding programs to transfer their beta carotene–producing genes into other rice varieties for use by poor farmers in developing countries. But the thicket of patent protection and intellectual property rights agreements seemed impenetrable, and their early attempts to negotiate free access to Golden Rice failed. This was particularly frustrating for Potrykus, whose experiences as an eleven-year-old hungry refugee from East Germany left him with a life-long humanitarian commitment to conduct research focused on alleviating malnutrition.

But then something unusual and unprecedented occurred, at least for the biotechnology industries: Monsanto offered to provide its proprietary components of Golden Rice and its own beta carotene–enriched canola and mustard crops royalty free. The corporate megalithic monster suddenly had morphed into the good guy.

Monsanto's home offices are housed in two upscale communities outside St. Louis, Missouri, corporate headquarters in Creve Coeur and the laboratories where genetic engineering happens across town in Chesterfield. Creve Coeur has a university feeling to it, with its conference center surrounded by beautifully landscaped grounds, and its meeting rooms sporting visionary names such as Adventure, Believe, Create, Discovery, and Explore. I met there with several Monsanto employees in the spring of 2001, and afterward was given a tour of Monsanto's Chesterfield Village Research Center, a six-story complex with two acres of greenhouses on the roof. Containing two hundred and fifty laboratories, it had cost $150 million to construct in the mid-1980s, a price tag that made it the biggest construction project in the state up to that time.

My visit was timely, because Monsanto had startled the biotechnology community in October 2000 by agreeing to donate their relevant beta carotene technologies for humanitarian purposes. Their intellectual property contribution to Potrykus and Beyer's Golden Rice research program was a compound called a promoter that intensifies the expression of the daffodil genes imported into rice, a step critical to increasing beta carotene production to nutritionally adequate levels.

The complicated negotiations between Potrykus and Beyer on one side and the biotechnology industry on the other had been stalled, but the Monsanto announcement jump-started the process. Potrykus and Beyer used the Monsanto commitment as leverage, and the other companies and inventors fell in line within a few months by similarly donating royalty-free rights to the materials and technologies they controlled. The agreement that ensued allowed Potrykus and Beyer to retain the rights to share the technology with producers in developing countries who earn less than $10,000 U.S. annually, a group that includes almost all third world farmers.

The transfer of the Golden Rice technology into developing countries has been dubbed the Humanitarian Project and has its own board of directors, including Potrykus. Its key components include the participation of the multinational company Syngenta, funding from the Rockefeller Foundation, and the breeding and transfer of new varieties to farmers through the International Rice Research Institute in the Philippines. Syngenta is providing plant-breeding expertise and conducting the biosafety and nutritional studies required by regulatory authorities in various nations. The International Rice Research Institute is using funding from the Rockefeller Foundation to develop Golden Rice breeding stocks, to be distributed free of charge to farmers in developing countries. These varieties will be self-pollinating and true-breeding, so that growers can save a portion of the seeds annually and use them to grow subsequent crops. Thus the benefits of Golden Rice will be self-perpetuating, and poor farmers will not have to purchase seed annually.

Monsanto's contribution was key to the development of Golden Rice and to facilitating its humanitarian transfer to poor farmers, but the company's own in-house research has focused on oilseeds, particularly the closely related canola, rape, and mustard plants. Mustard is poised to be the first of Monsanto's beta carotene–enriched oilseed crops to be fully integrated into a third world development program, in India. Mustard oil is second only to peanut oil as a food-preparation and cooking product in Indian kitchens. Nicknamed "Golden Mustard" although it is actually deep red, it has an excellent chance of relieving vitamin A deficiencies. Less than one teaspoon of Golden Mustard oil provides the daily minimum amount of beta carotene recommended to ensure adequate vitamin A production in a child.

Monsanto has forged agreements to transfer its beta carotene–producing genes into local Indian mustard varieties in cooperation with the Indian government, the U.S. Agency for International Development, and the nonprofit Indian Tata Energy Research Institute (TERI). The next steps are to cross Monsanto's genes into local Indian mustard varieties, to carry out full environmental and human safety assessments, and to conduct clinical trials to determine whether the enriched oil derived from the new plants actually does increase vitamin A levels. All of this work will be done in India, and the final product is still a few years away from being widely available.

Monsanto's role as donor exemplifies the type of interactions that the biotechnology industry is promoting in developing countries. I talked with Robert Horsch, vice president of product and technology cooperation, about Monsanto's donation of Golden Mustard technology to India and its evolving relationships with developing countries. Horsch is well qualified to discuss intellectual property rights; his own research achievements were recognized in 1998 when he was awarded the National Medal of Technology for his work in plant genetic engineering. Horsch described Monsanto's role as sharing knowledge rather than selling products: "We're making available to TERI intellectual property. It's a multipart collaboration. In the end it will be a public-sector product, we won't make it, certify it, or sell it, but we will make available what we know. We don't have any expectations of royalties; we're not going to make a business out of it."

Why, we may well ask, did Monsanto decide to donate its potentially lucrative beta carotene technologies? The motives are multiple, and as is the case with most philanthropic activities, involve a blend of self-interest and altruism.

Monsanto's beta carotene crops do have some potential for profit outside of the international development sector. Monsanto hopes to recoup some of its investment in biotechnology research by retaining marketing rights in developed countries. Not all farmers are poor, and more prosperous producers are expected to purchase the enriched seeds when the demand for nutritionally enhanced products increases. Moreover, health-conscious consumers in developed countries are very likely to buy enriched foods, especially foods enhanced with Vitamin A or beta carotene, because these are anti-oxidant compounds that are thought to protect against cancer.

The donation of Golden Mustard to India is also likely, as Horsch pointed out, to help in the eventual development of other markets as third world economies improve: "There is a growing belief that humanitarian efforts allow movement beyond subsistence into real economic activities that will generate new customers and markets for us. This is hard to prove but we believe that this is a good long-term business development strategy for Monsanto."

Monsanto was probably also motivated by a desire to mute or silence critics of its biotechnology empire. Its current philanthropic bent reflects a changed public relations strategy to position Monsanto as a company with a conscience, and also a determination to begin moving products even if profits are not immediately forthcoming.

I spoke with David Kryder from the International Service for the Acquisition of Agri-biotech Applications (ISAAA) about the intersection of biotechnology and philanthropy. ISAAA is located at Cornell University in Ithaca, New York, and is one of many not-for-profit organizations that have sprung up around the bioengineering community with the aim of facilitating agricultural biotechnology in developing countries. Kryder was formerly in charge of licensing agreements for Pioneer Hi-Bred International in Des Moines, Iowa, and is now senior specialist for intellectual property transfer at ISAAA. He was one of the authors of an ISAAA publication released in 2000 and funded by the Rockefeller Foundation that described various intellectual property options that might improve biotechnology transfer to the third world.

Kryder thinks that biotechnology companies are suffering from their early history of bad public relations, and that engaging in philanthropy is a good way both to overcome their poor reputation and to get their products on the market. Corporate donations, he says, combine good business sense with generosity: "Nobody's making any money on GM products right now. If you were in that business and you'd invested hundreds of millions of dollars in research, wouldn't you try to do something that would market your product? You'd find some way to promote GM products; only a fool would not. Is there some self-promotion in this? Of course, but does that mean this is inherently evil or some slimy, subversive, snake-eyed corporate thing? I don't think so."

Monsanto's decision to donate biotechnology has also had a positive effect on its employees, many of whom chose to work in the field because they want to improve the human condition through genetic engineering. They are pleased that the transfer of beta carotene technology to developing countries will reduce malnutrition and believe that the company is behaving admirably. As Cherian George put it, "I think Monsanto is leading the way; that's the way I see it. There hasn't been anything like this before. Most of the first products were for farmer productivity, but this is the first one that has a very high impact on society. We're hearing a lot of nonsense and criticism from people who don't even know what the facts are. That will always be there. We're just giving the technology to people who want it. We're not expecting the activists or anybody to say anything positive about Monsanto. This is what is right, let me put it that way."

Monsanto ventured into corporate philanthropy at the same time that a vigorous debate was beginning about the role of transgenic crops in third world agriculture. On one side of this continuing debate are the cynics, who decry the downloading of biotechnology to developing countries as nothing but a manifestation of corporate greed. On the other side are the proponents, who extol these new crops as important contributions to human welfare. And underlying this debate is a broader intellectual conflict, between those who believe that technology has great potential to feed the third world and those who advocate traditional indigenous agricultural methods and small-scale, sustainable farming as the solution to global hunger.

The two sides that have squared off in this battle fundamentally agree on the dire problems faced by third world farmers but differ dramatically in their proposed solutions. The established international development elites clearly approve of biotechnology. They work in a mix of academically distinguished universities, prestigious privately funded think tanks, and well respected humanitarian foundations, whose endowments exceed the budgets of many of the countries they serve. On the other side are groups of protesters who passionately believe in simple low-technology solutions rather than in high-technology transgenic crops.

Robert Paarlberg is at the forefront of the academic analysts who favor the use of biotechnology to help improve agriculture in the developing world. Paarlberg is a professor of political science at Wellesley College and an associate at the Weatherhead Center for International Affairs at Harvard University. We met on a blustery November day in 2000 at his Harvard Office, under the impassive gaze of dozens of exotic masks mounted on the walls, souvenirs of his extensive international travels.

The Weatherhead Center is the quintessential academic institution, a place for scholars in the Boston area to attend seminars, criticize each other's working papers, and engage in discussions that have considerable influence on the shaping of government policy. Paarlberg has written key articles for influential journals such as *Foreign Affairs*, and in 2001 he published a book about the policy dilemmas facing developing countries that want to use biotechnology to improve agriculture.

Paarlberg and most other scholars view biotechnology as offering ways of relieving third world poverty that are more effective than those previously recommended. Both the development agencies and the protest communities agree that the much-touted Green Revolution has not lived up to expectations. The Green Revolution began with the development of higher-yielding wheat and rice varieties during the 1950s and 1960s, funded primarily by the Ford and Rockefeller Foundations. These new varieties were beneficial for some, but they require a lot of water and relatively high fertilizer and pesticide inputs to succeed.

The Green Revolution has been criticized for pushing agriculture in third world countries toward the high-input farming characteristic of developed nations. Poor farmers often cannot afford to grow the crops that drive the Green Revolution, and many who did grow the new crops ended up with lower yields after abandoning the traditional crop varieties that were better suited to regional conditions and that had been selected over thousands of years for resistance to local pests and diseases.

Pests, diseases, low nutrients in the soil, and lack of water are all management problems that Paarlberg thinks transgenic crops can solve. GM seeds can be engineered to enhance productivity without the application of costly pesticides and fertilizers and without expensive irrigation, and as Paarlberg says, "Transgenic crops are less complicated. If you have smart seeds, you don't need to have farmers with the same instant upgrade of knowledge. Biotechnology is actually easier for low-resource, illiterate farmers to adopt."

Scholars like Paarlberg argue that pest and disease impact is more severe in third world countries, where many plant pests and diseases originated and where their diversity remains high. He gives examples such as the 15 to 45 percent of maize crops lost annually to a diverse array of insects in Kenya; that loss could be diminished by a switch to Bt maize, which thrives without the application of expensive chemical pesticides. In Mexico, small-scale potato farmers who now suffer substantial losses from plant viruses could increase their yields by switching from conventional varieties to transgenic virus-resistant potatoes. Crops bioengineered to be drought-resistant and salt-tolerant could also allow farmers to make use of land that was previously unproductive.

The international development institutions downplay the potentially negative impacts of GM crops. They say that conventional or Green Revolution agriculture does more environmental damage than biotechnology will cause, and that transgenic crops are more likely to improve rather than harm third world ecosystems. For example, GM crops can reduce pesticide use, and that will be an environmental plus in the third world, where pesticides, many so toxic that they are banned in developed countries, have been extensively overused.

Similarly, the potential impact on biodiversity of bioengineered genes that might jump to native plants is minor in the face of the intrusion of agriculture into diminishing natural habitats. Any improvements in yield from transgenic crops would reduce the need to expand farming into threatened ecosystems to increase production. Food safety is also of little concern in the third world, where spoilage and disease transmission are much more serious problems than the remote possibility of allergic reactions to GM products. Food availability and its price are of paramount importance, and the enhanced nutritional quality of transgenic crops holds out the promise of a net improvement in both the quality and the amount of food.

Proponents of biotechnology from the international development community also are not overly concerned that multinational companies will exploit the third world. They point to agreements between the development community and corporations that have reached a middle ground between profit and humanitarian aid. Paarlberg himself is more worried that business will not become involved in the earth's poorest regions because of low profits and political instability.

For Paarlberg, multinational companies are not responsible for the persistence of poverty around the globe, and he thinks protest against biotechnology in the third world is often rooted not in disagreements about scientific issues but in lingering distrust of large corporations: "I think the aversion to multinational firms is an understandable rallying cry, but I don't think it's an accurate diagnosis of what's keeping poor people poor. You never quite know if the opponents of biotechnology are saying no to Bt cotton in India because they are not sure what will be done to biota in the soil by the decaying crops or if they are just opposed to any technology brought into India by U.S multinational companies like Monsanto."

■■■ Third world opposition to genetically modified crops has often been shrill and accusatory. One of the more extreme anti-GM campaigners is Vandana Shiva of India, a writer and the director of the Research Foundation for Science, Technology, and Natural

Resource Policy. Her harsh criticism of governments in developed countries and multinational corporations conveys the essence of the radical conspiracy theories that hover at the fringes of antibiotechnology activism.

In her interviews and writings she denounces crops that reduce vitamin A deficiencies as "a blind approach to blindness control being used as a Trojan Horse to push genetically engineered crops and foods. The real issue for both people and nature is the extent to which control over seeds and other genetic materials is becoming increasingly concentrated in the hands of those whose only interest is profits. The intellectual property rights regimes being put into place through GATT set the stage for foreign corporations to gain a total monopoly control of our food production by displacing traditional seed varieties with patented hybrids."

Strong rhetoric, but mild compared to that elicited by the Orissa incident. In 1999 over ten million people were left homeless following a cyclone that devastated much of the eastern coastal state of Orissa, India. Humanitarian agencies mobilized emergency food shipments, including one shipload from the United States laden with a highly nutritious corn-soybean mixture provided by the U.S. Agency for International Development (AID). Shiva had a sample tested. She found traces of GM crops in the mixture, which is not surprising because much of American corn and soy planted at that time was genetically modified.

Shiva immediately issued a press release calling for the withdrawal of the donated food, and accused AID of using the storm victims as unwitting subjects of a GM experiment: "The United States has been using the Orissa victims as guinea pigs for GM products which have been rejected by consumers in the North, especially Europe. We call on the government of India and the state government of Orissa to immediately withdraw the corn-soya blend from distribution." No matter that the same corn and soy products had been used in many U.S. food products for several years by then, without incident, and that some people may have starved without this genetically modified food.

The most extreme anti-GM incident on record also involved AID and international issues, but took place in the United States. In 1991, AID signed a cooperative agreement with the Institute of International Agriculture at Michigan State University to serve as the lead organization in a consortium of U.S. institutions devoted to implementing agricultural biotechnology in the third world. The Michigan group soon caught the attention of the violent anti-GM fringe. On 31 December 1999 the institute was the target of a successful arson attack by a shadowy group calling itself the Earth Liberation Front. The institute's offices and all its records were completely destroyed, but fortunately no one was injured.

There are, of course, more moderate opposition voices that focus less on corporate conspiracy and more on the potential of alternatives to transgenic crops to improve nutrition and health in developing nations. These more rational activists begrudgingly admit that crops such as Golden Rice might indeed alleviate vitamin A deficiencies, but argue that there are better ways of combating hunger and malnutrition that are more compatible with the resources available to third world countries.

One of the more balanced and measured documents championing a home-grown approach was the April 2000 United Kingdom report *Biotech—The Next Generation* by Sue Mayer and Sue Dibb. They wrote, "The importance of GM solutions to micronutrient deficiencies in developing countries should not be overplayed. It has not been the absence of solutions that has hindered progress but political, economic, cultural and social factors. GM crops will not eliminate these factors, although a focus on GM could divert resources inappropriately . . . Public policy initiatives and public expenditure should be targeted at encouraging balanced diets and healthy lifestyles rather than supporting developments in the area of nutritionally altered GM foods."

They point out, correctly, that there are other ways of supplying vitamin A. There are animal-based sources of vitamin A, especially egg yolk, chicken, milk, and butter; and beta carotene is present in many fruits and leafy vegetables, including spinach, carrots, pumpkins, and mangoes. Dibb and Mayer argue that aid programs would better serve the third world by encouraging the cultivation of those resources than by imposing a reliance on genetic engineering to introduce what nature can provide. Thus giving poor farmers chickens, milk cows, and vegetable seeds might accomplish the same objectives as giving them golden crops, with the further benefit of increasing food security, improving nutrition, and diversifying diets.

It is difficult to argue with this idyllic view of farming in a third world rural community. The images are pastoral and compelling: small plots containing diverse vegetables, chickens pecking at insects throughout the village, a communal milk cow, farmers in their rice paddies carefully tending varieties passed down for generations by their ancestors. Indeed, some regions of developing countries are like that, with enough economic stability and a sufficient land base for home gardening and locally grown foods to be important factors in the maintenance of a diverse, balanced diet and good health.

But many places in the third world are not like that, especially dense urban slums where millions of people are crammed into shanties and shacks with no land on which to grow crops or raise chickens. Over eight hundred million people suffer from hunger and malnutrition, the 15 percent of the world's population who consume less than two thousand calories a day. For them, the extent of food shortages and malnutrition is so great and the available resources so limited that simple backyard solutions alone are not enough. Broader-scale increases in crop production are necessary to reduce food shortages for these starving hundreds of millions.

■ Whether conducted in moderate or accusatory terms, the debate concerning the role of transgenic crops in solving hunger and nutritional problems in developing countries has been between those who want to avoid technology in favor of simpler solutions and those who strive for further technological breakthroughs. Transgenic crops may be such a breakthrough, and may have the potential to extend productivity and enhance nutrition in societies that lack the means to afford the inputs necessary for Green Revolution technology to succeed.

Both low- and high-technology approaches have merit, but poverty-stricken nations need to develop their own perspectives and policies about the balance between transgenic crops and small-scale indigenous agriculture. Like industrialized societies, they must consider environmental risks, food safety, intellectual property rights, and economics as they assess the known or potential benefits of biotechnology. But the stakes are considerably higher in the third world; hunger is so prevalent in these countries that their decisions often reflect an increased willingness to accept risk in order to obtain more and more nutritious food.

Whatever decisions third world countries make about transgenic crops, the programs they adopt must be self-sustaining and capable of being applied locally. Harvard's Paarlberg is critical of all attempts, whether by proponents or critics of GM crops, to impose industrialized values on impoverished developing nations. He thinks that the most sustainable approaches for countries that choose to pursue biotechnological responses to poverty

and malnutrition will be ones in which the infrastructure to produce and regulate transgenic crops is developed at the local level.
As Paarlberg told me:

"Tragically, the leading players in this global GM food fight—
U.S.-based industry advocates on the one hand and European
consumers and environmentalists on the other—simply do not re-
liably represent the interests of farmers or consumers in poor
countries . . . The subsequent public debate naturally deteriorates
into a grudge match between aggressive corporations and their
most confrontational NGO [nongovernment organization] ad-
versaries. Breaking that paralysis will require courageous leadership, especially from policymakers in developing countries. These
leaders need to carve out a greater measure of independence
from the GM food debate in Europe and the United States. New
investments in locally generated technology represent not just a
path to sustainable food security for the rural poor in these countries. In today's knowledge-driven world such investments are increasingly the key to independence itself."

Paarlberg's vision of developing countries investing in their
own homegrown biotechnology is admirable, but turning it into
reality will take time and money. Malnutrition and hunger are
pressing realities, widespread and increasing. More immediate solutions require that corporations continue to transfer technology
at little or no cost.

The behavior of Monsanto and the other biotechnology multinationals is often described as being either enlightened or exploitive, humanitarian or profit-seeking, but corporations can
certainly have mixed motives. Monsanto's position is that profit
and philanthropy can go together. But whatever the company's
intentions, its willingness to move some nutrient-enhanced transgenic crops into developing countries royalty free is a positive

step. Consumers' anger at Monsanto's past practices may be well deserved, but if the company truly is changing, then opposition to its entry into third world markets may hinder the alleviation of food shortages and nutritional deficiencies.

The word "profit" is defined as "an advantageous gain or return," and brings to mind the phrase "profit at the expense of . . ." The issue for Monsanto and its opponents is to reach a point where both the company and developing countries benefit, at no one's expense. For both Monsanto as a business and the developing world as a disadvantaged region, that point will have to balance corporate profits with humanitarian impulses.

The equilibrium point is not the same in rich and poor countries. In the developed world, Monsanto's initial profits from genetically modified crops were considerable, although they now have decreased somewhat; the benefits to farmers and consumers were moderate from the first generation of insect-resistant and herbicide-tolerant plants; and the health and environmental risks were nonexistent to extreme, depending on whom you ask. In developing countries, Monsanto's current profits are small and the possibility of increased income in the near future is slim, but the potential benefits from nutritionally enhanced crops are substantial. The risks are similarly dependent on perspective and opinion, but seem less immediately worrisome when weighed against urgent needs.

Only time and experience will tell whether nutrition-enhanced transgenic crops will be successful. Monsanto, however, is betting much of its long-term biotechnology future on these crops. In addition to products enhanced with beta carotene, researchers are working on rice varieties rich in iron and vitamin E, vegetables with healthier fatty acid profiles, fruits with enhanced levels of the phytonutrients thought to protect against cancer, crops such as transgenic peanuts with reduced levels of allergy-triggering proteins, and produce that can be stored for prolonged periods of time.

Monsanto may have been stunned by the extent of the international opposition to its flagship role in genetic engineering, but its current corporate rallying cries reflect a growing optimism that nutritionally enhanced foods will revive the reputation of biotechnology. The signs and banners at Monsanto's Chesterfield research and visitors' center convey this spirit: "We're a New Company," "Abundant Food and a Healthy Environment," and "Tomorrow's Harvest." Monsanto's view is that the good of mankind and corporate success are synonymous, and the company's future rests on the expectation that the next generation of genetically modified crops will inspire the public to share that vision.

The present debate over genetically modified foods is more indicative of the poorly planned use of an immature technology than of a failure of the technology itself.

Robert Carlson, winner of the silver award for his essay "The World in 2050," The Economist/Shell World in 2050 Essay Competition, 2000

Risks Real or Imagined

Immersing myself in the subculture of genetic engineering has been both fascinating and unsettling. The science itself is novel and gripping, expanding our understanding of the most basic aspects of living organisms and building the foundation for the creation of a host of lucrative products. Yet our developing relationship with biotechnology has been rocky, revealing some serious flaws in the way we deal with scientific issues and manage new technologies.

The benefits of genetically modified crops are not as great as proponents would lead us to believe, and the risks are not nearly as intractable and harmful as the critics maintain. Biotechnology provides useful tools, but it is not a solution for the fundamental problems underlying contemporary agriculture, nor will it lead to tragic meltdowns for environmental protection or human health. The reality is more complex than either proponents or opponents admit.

The most distressing aspect of my travels in the genetically modified zone was discovering the intense balkanization and profoundly antithetical points of view that characterize biotechnology debates. Most people and organizations I encountered lacked any willingness to consider opposing opinions, holding firmly to their own views. Be they corporations or environmental groups, organic farmers or regulators, consumers or lawyers, interest groups too often perceive issues about genetic modification from behind the walls of their own intellectual fortresses, with bulwarks maintained by strong biases and slanted perspectives.

Industry sees only the advantages of GM crops and has been unwilling to acknowledge any negative side effects, no matter how inconsequential they may be relative to the benefits. Consumer groups panic at the potential for deadly allergic reactions in spite of the absence of any such incidents, and disregard the potential nutritional benefits of GM foods. Organic farmers perceive transgenic crops exclusively as a threat to their production systems and do not acknowledge their potential for reducing pesticide use and improving the environmental record of conventional farming. Regulators respond in see-saw fashion to the intense lobbying of pro and con pressure groups, losing sight of their true mission, to act in the public interest.

All sides fail to deal with the scientific data objectively, and tend to create biotechnology legends from small bits of unconfirmed information. The broader public has been besieged by sound bites and public relations hype rather than exposed to comprehensive and informed debate and dialogue. Many of the advocates and opponents of genetic engineering are thoughtful, reflective, and balanced in private discussions, but their public statements and their organizational positions too often are inflexible and factually imprecise, designed to manipulate rather than inform public opinion.

Spin and hyperbole are not, however, unique to arguments about genetic engineering. They permeate discussions of many

scientific issues that are generating increasing demands on societal decision making, including global warming, new reproduc-
tive technologies, and the significance of diminished biodiversity.
We need to take a more balanced view of all such issues, judi-
ciously weighing the benefits we can derive from scientific discoveries against the risks they pose. To find that equilibrium point
between benefit and risk, we need to repudiate the extreme positions and intense lobbying of both advocates and critics. In the
case of biotechnology and similar issues, we need to explore how
interest groups might begin moving toward that middle ground
where the public's best interests are found.

Corporations have always been protective of their technologies, sensitive to criticism, and overly quick to respond aggressively to any threats to their freedom to operate and their
ability to turn a profit. But the companies that create and market
genetically modified crops have been unusually self-serving.

They have lobbied hard for minimal regulation, presented a
united front in refusing to acknowledge the possibility of even
slight environmental side effects, and vehemently denied that
food derived from GM crops could in any way be unsafe. This
corporate stance may turn out to be scientifically correct, but it
has contributed to an unprecedented public outcry about transgenic crops that could have been headed off early on by a more
flexible, respectful, and accommodating attitude toward critics.

So far the industry has been unwilling to address public
concerns by providing complete data concerning health and environmental impacts or by supporting the enhanced regulatory
oversight that might improve public confidence in biotechnology.
This arrogant approach has backfired; strong opposition in continental Europe and in the United Kingdom has resulted in vir-

tually no European markets for GM crops or foods. North Americans have been more receptive, but opposition campaigns are taking their toll and industry is beginning to feel the pressure exerted by shrill but clever and effective anti-GM environmental groups.

Considerable damage has been done to public confidence by corporate posturing about genetic engineering, but the industry can still redeem itself. What the public sees and mistrusts is expensive, heavy-handed, manipulative, and one-sided corporate lobbying. But behind the industrial barricades are many dedicated, committed researchers and marketers who believe in the benefits of their work and have confidence in the quality and safety of their products.

The inconsistency between the public statements and the private opinions of corporate researchers reflects the movement of scientists from basic, independent, university- and government-housed science into the more restricted and secretive world of industry. This transfer of talent and effort has led to an accelerating output and an increasing diversity of products, but it has also had profound implications for how science is conducted and perceived. Academic and industrial scientists do benefit from the cross-fertilization resulting from town and gown cooperation, and many of our most significant medical and agricultural products can be attributed to this fusing of basic and applied science. But the growing erosion of traditional scientific openness under the corporate mandate for secrecy has also been detrimental to the public good.

These are not intractable problems, and they can be addressed in many ways. First, corporations need to relax constraints on their scientists and allow them to openly express diverse and even critical opinions. The many corporate scientists I talked with perceived the benefits and risks of GM crops and discussed those privately with a balance not evident in their public statements. The industry's credibility would be enhanced if these employees were permitted to speak openly to the public. Advocacy rings truest when it is based on realistic expectations rather than exaggerated claims.

Industry also should shift research about environmental safety and human health to the public sector, and be prepared to fund it without restrictions. Biotechnology firms are beginning to understand the fundamental importance of yielding control of environmental and health research to independent bodies, and the benefits of that approach were apparent during the monarch butterfly controversy. In that case, a consortium of companies provided no-strings-attached funding to American and Canadian universities and state and federal government laboratories to examine the impact of Bt crops on these and other butterflies. The results suggested little impact and, more significant, reassured a skittish public, so that this particular issue is no longer of major concern. The industry should encourage rigorous analysis by independent parties when similar issues about GM crops arise, and be prepared to withdraw products that do not pass muster.

Another useful approach for industry to pursue would be to insist on tougher regulatory oversight. The corporate stance that only light government supervision is needed for GM crops and the foods produced from them may be scientifically correct for transgenic plants, but this attitude has alienated a distrustful public and proven to be a poor tactic in the bid to build consumer confidence.

The first recombinant DNA researchers chose voluntarily to set an unusually high safety standard for their research, and then supported legislation to enforce it. Time and experience proved the research safe, and it now proceeds with few restrictions. The tough regulations that initially were suggested and imposed by the researchers themselves defused the public's concerns and engendered confidence, and the biotechnology industry would be well advised to follow the same path.

■■■ The defensiveness and intractability of the biotechnology industry did not develop in a vacuum. It grew and hardened in reaction to unreasonable, irrational, and emotional attacks by environmental and consumer groups. It is easy and tempting to portray industry as the bad guy, but opponents of GM crops have been equally guilty of polarizing biotechnology issues. Their extreme rhetoric and attention-grabbing tactics have helped turn what should have been a reasoned public debate into an exercise in mud-slinging and name-calling.

The rhetoric of the critics has matched industry hype by exaggerating risks, warping facts, and latching on to bad science. Protesters cite poorly replicated laboratory studies involving the administration of abnormally high doses to test animals when they proclaim the dangers of transgenic crops to non-target organisms, but they ignore field studies indicating little or no impact. The stories about the jumping of antibiotic-resistant genes from crops to bacteria that circulate through the linked Web sites and press releases of GM opponents are not backed up with publicly accessible data. Greenpeace's antibiotechnology icon FrankenTony and its Frankenfoods wordplay are fun and attention-grabbing, but there is no evidence that Kellogg's products or other transgenic foods have actually harmed anyone. Passions inflamed by these and other misrepresentations of biotechnology have led to fire bombings of university buildings, clandestine nighttime uprootings of GM crops from farmers' fields, and claims that GM food donated for disaster relief in the third world was tainted.

Critics of transgenic crops have lost as much credibility as industry through these one-sided and sometimes violent attacks, but many protesters feel that they have no choice. The intensity of their tactics has been fueled by a sense of powerlessness, heightened by the perception that government regulators favor industry, the belief that corporations are profit-driven entities lacking ethics and responsibility, and a fundamental mistrust of government and industry assurances that all is safe in the genetically engineered world.

The protest groups have raised some important issues about GM crops, but their excessive zeal has diminished their trustworthiness as much as the corporations' arrogance has diminished theirs. The most unfortunate collateral damage resulting from this destructive debate has been the loss of perspective on what appropriate scrutiny of biotechnology should involve. The posturing and the accusations batted back and forth by extremists on both sides have obscured the central issue that needs to be probed: Which gains are worth which risks?

The fact is that we have not yet decided what level of side effects from GM crops are worth the benefits. The negative impacts found so far range from nonexistent to slight to moderate; there have been no outright disasters. There is no evidence of health risks from any current GM food. There have been minor and sporadic impacts on non-target insects, vertebrates, and other organisms in the field, and there is some reason to be concerned that herbicide-resistant genes that jump to non-GM weeds or crops will transform them into super-weeds, immune to many herbicides. None of these effects are any worse than those caused by conventional agricultural practices, and the reduced pesticide use possible with GM crops in many instances has to be considered a plus for the environment and human health.

More testing may reveal more serious problems, and new crops may cause more serious consequences than currently commercialized transgenic plants. For some observers even the level of impact found to date is too severe, while for others the benefits of GM crops would justify considerably more extensive side effects. There is no one correct point of view, but we need to delineate our communal comfort zone. Doing so is the legal role and the responsibility of government regulators, but unfortunately U.S. and Canadian regulatory policies have done little to bolster public confidence. It is in the regulatory arena, where industry and opponents meet to influence the rules, that decisive and fairminded action could be taken to counter the public's concerns about GM products.

The job of government regulator may be the least respected and most highly pressured in the developed world. The typical regulator labors in an unstable and tense workplace, besieged by lobbyists who are strongly motivated by self-interest, confused by fluctuating agendas set out by elected political masters beholden to campaign donors, supervised by appointed officials who often have little expertise in the areas they oversee, and pressed to make decisions on the basis of what invariably are too little data.

The regulator's role is simple in concept but remarkably complex in execution: to mediate between the corporate drive for profitable products and the public's need to be protected from any damaging side effects those products might cause. Industry's position on regulatory control is predictable: the corporations favor decreased oversight. Businesses argue that their corporate sense of responsibility provides sufficient protection for consumers, that stakeholders are the best arbitrators of safety, and that regulation is too expensive, delays or prevents the marketing of their products, and diminishes profits and thus economic growth if too heavily imposed.

The position of consumer and environmental groups can be just as predictable: they are often dissatisfied with even the most rigorous regulations. Critics of the regulatory system fear that negative impacts have been hidden or misrepresented by industry and that unknown or unrevealed disadvantages of GM products could be injurious to humans and the environment.

These opposing forces drive regulatory decisions into the compromise position of providing the minimal amount of protection needed to prevent outright disasters but not guarding against more subtle or long-term impacts. So far, North American regulators have done enough to contain the most serious potential consequences of GM crops but not enough to dispel the public's uncertainty about safety, while industry and its opponents continue to pressure the regulators to do more or less than they have done.

To see how difficult the regulators' task can be, consider the issue of possible allergic reactions to transgenic foods. In the United States, according to current FDA policy, food products must be labeled if they contain proven allergens. Scientists from Pioneer Hi-Bred created a variety of genetically modified soybeans with proteins originating from Brazil nuts, with the result that anyone allergic to Brazil nuts might have had a serious or fatal reaction after eating the bioengineered soybeans. Tests conducted long before the GM soybeans were ready for marketing showed that such reactions could occur, and Pioneer chose not to market the product rather than submit to the labeling requirement.

This example is frequently cited by industry to indicate that the regulatory process for transgenic crops is working, but opponents cite the StarLink incident as proving that GM regulation has failed. As you will recall, the StarLink variety of GM corn was approved for use only as animal feed because all of its proteins were not quickly degraded through digestion, although that does not necessarily mean that the corn would cause allergic reactions in humans. Some StarLink corn made its way into corn-based products sold to humans. Anti-GM forces were furious and many consumers were convinced that a lax regulatory system had put the public at risk. Industry countered by arguing that the StarLink should have been approved for human use in the first place, a point supported by the lack of evidence that StarLink Bt corn can damage human health.

What is most illuminating about these two incidents is that the perception persists that people are endangered by potential allergens in GM food, even though no such effect has been proven. Ironies abound: For instance, peanuts are known allergens that can induce fatal reactions, and foods must be labeled if they contain peanuts. Yet there is no public outcry in favor of banning peanuts from fields and grocery shelves, and corporations have not let peanut labeling requirements inhibit the marketing of their products, as Pioneer did with GM soybeans. In the case of the StarLink products, the deluge of media reports left the impression that the risk was considerable rather than remote and unproven, and the result was a $100-million buyback of a product based on undemonstrated negative health effects.

Regulators should be admired for their accomplishments, given the intense lobbying pressures they face and the difficulties posed by the fears that drive public opinion. Nevertheless, the regulatory system is flawed, and the most ironic aspect of biotechnology regulation is that the lax regimes encouraged by industry do not serve business itself well. The mild regulation that is consistent with corporate wishes merely fuels public fears about the safety of GM crops. "Regulation Lite" ignores the fundamental reality that consumers' perception of safety is as important as safety itself.

More publicly acceptable regulation of agricultural biotechnology is scientifically and economically feasible. We all, corporations included, would benefit from a broadly based and bold move by industry to support increased and independent testing of GM products for environmental and health safety as well as the labeling of foods containing genetically modified components. Conventional wisdom in corporate boardrooms has never favored increased oversight, independent evaluations, or clear product labeling. But given the intense and increasing public concerns about transgenic agriculture, the conventional wisdom is not serving the GM industry well.

At a minimum, the biotechnology industry should work with consumer and environmental groups to develop more stringent regulatory standards for GM products based on a few simple principles. First of all, each new type of genetically modified food should be tested individually for safety and environmental impact; the old standard of substantial equivalence should be discarded, at least until consumer confidence is established. Few GM foods would fail a rigorous battery of tests, and the relatively small expense and slight delay in getting the products to market would be well worth the gain in consumer confidence.

A second operative principle for biotechnology regulators should be that environmental tests be conducted before products reach the market, and that the tests themselves go beyond the standard laboratory analyses of effects on non-target organisms that are used to register new pesticides. Additional field tests should be run on the specific organisms most at risk from any new transgenic crops. Had monarch butterflies been the subject of testing before, rather than after, Bt corn reached the market, there would have been much less confusion and public outcry about the effects of the corn on these butterflies. We particularly need to increase research on the extent to which genes jump from GM crops to wild plants and on how often cultivated GM crops become weeds. Both phenomena occur, but without further research the impact of gene-jumping incidents and weedy crops on ecosystems cannot be established and the appropriate management responses cannot be formulated.

A third and long overdue regulatory policy should be to have independent organizations conduct safety research rather than to rely on data generated by the industry that makes the products. Industry does have a stake in assuring the safety of its products, and in that regard is a more reliable source of information than critics perceive. Nevertheless, corporate studies will never be considered unbiased, and the obvious way to avoid perceptions of bias is to have third parties conduct environmental impact and human health testing at universities, government laboratories, or nonprofit research institutions. Government regulators still need to decide what level of safety is sufficient, but at least the distrust of research conducted by industrial laboratories can be eliminated.

A final step that the regulators and the biotechnology industry should consider is reversing their opposition to mandatory labeling for GM foods. Instead of saying, "since our products are safe, why label them?" biotechnology companies could develop a different marketing strategy, one accentuating the positive: "Let's market our GM foods with confidence in their safety and pride in the products; since GM foods are an improvement over conventional foods, and are safe, we will advertise these benefits through labeling." Such labels could read "This product comes from genetically improved plants that have been designed to protect crops from pests and reduce pesticide use," or "This product was made from genetically modified crops that increase yields and thereby help reduce world hunger."

Clearly, product labeling, stricter testing, independent laboratories, and full disclosure of scientific studies would all increase consumers' confidence in transgenic crops and foods. But to date the biotechnology industry has argued effectively with regulators that none of these is necessary, and, in North America anyway, governments have leaned more toward boosting biotechnology than policing it. Yet in the future a more appropriate balance must be reached between encouraging industry and furnishing regulatory reassurance if consumers are to believe that they are well protected from risks imagined or real.

Debates about genetically modified crops have often brought into sharper focus some basic characteristics of contemporary agriculture. Attitudes about farming today often are polarized. On one side are the conventional growers, who use synthetic pesticides and fertilizers and often grow genetically modified crops, practices considered by some to be hard on the environment and to produce residue-tainted food. On the other side are the organic farmers, who use methods that have a gentler impact on the environment and that are perceived as yielding more healthful food, but who are criticized for not producing high enough yields to feed the many hungry people in the world.

These two groups are antagonistic, neither conceding that the other has a credible approach to agriculture. Proponents of organic farming passionately oppose GM crops as inconsistent with their philosophy and resent the drift of bioengineered genes into their crops. Conventional growers ridicule organic farming as being little more than a hobby and not the way to produce the vast amounts of food needed to alleviate world hunger. In the developed world, the debate between these groups is often bitter, but at stake is primarily profit or loss and consumer preferences, since more than enough food is produced at affordable prices to feed its residents.

In developing countries there is more at stake, since decisions about farming practices can make the difference between subsistence and starvation. There also, public dialogue has become polarized. GM crops in third world countries are promoted as providing food with substantially increased nutritional content and crops with improved characteristics such as pest resistance, drought resistance, and salt tolerance. Opponents view GM crops in developing countries more cynically, as exploitive tools foisted on the third world by multinational corporations motivated only by profit.

I encountered many strong opinions during my exploration of attitudes about biotechnology, but none more passionate than those expressed by conventional or organic growers when asked about each other's farming practices, or by people concerned with third world issues when discussing the role of GM crops in developing countries. All sides were remarkably inflexible, almost always demarcating firm lines between acceptable and forbidden practices and refusing to acknowledge the validity of other groups' approaches. Rarely did I hear the more balanced view that sustainability may lie in diversity rather than in reliance on one system, or that reaching some middle ground between opposing opinions was possible and desirable.

Arguments about biotechnology that compare organic and conventional farming, or agriculture in developed versus developing countries, are based on the premise that one system is best and must predominate. It would be much more helpful to encourage a diversity of agricultural approaches, and add blends of biotechnological and traditional approaches to the more extreme recommendations that have characterized the transgenic debates so far. Is reaching such a middle ground possible, and can GM crops be planted on a wide scale without fouling organic approaches and corrupting traditional, indigenous crop varieties and agricultural practices?

The answer is yes, but only if all parties are willing to compromise and acknowledge the virtues of other people's methods. The opposition of the organic community to any transgenic crop may be based on valid concerns, but lack of respect for organic farming on the part of conventional growers has so intensified that opposition that it borders on the fanatic. Similarly, the dismissive attitude expressed by organic farmers in discussions of conventional farming has hardened positions on the other side, which makes dialogue difficult and finding a middle course nearly impossible.

A change in perspective from uncompromising criticism to respect for diverse agricultural approaches is critical if we are ever to improve food production throughout the world. From this simple change in attitude, we could then progress toward devising concrete strategies for achieving truly sustainable agriculture.

One area where compromise could function effectively is the certification of organic food. So far, organic producers and customers refuse to accept even a tiny amount of genetic drift from GM into organic crops, an attitude that is overly rigid and inconsistent with their acceptance of small pesticide residues in organic produce. Conventional growers for their part have been uncooperative in modifying their farming practices to minimize the possibility of such drift.

There certainly seems to be a potential middle ground here: The organic industry could certify foods as organic that have a minuscule proportion of accidentally introduced GM components, perhaps up to a tenth of 1 percent. Conventional growers could agree to minimize the possibility of drift by not planting bioengineered crops in regions with a high proportion of organic farms. They also could cover their harvested seeds securely when trucking them to market to prevent wind-blown volunteer GM seeds from germinating as weeds, and they could plant broad perimeter strips with conventional varieties, a good practice in any case because it minimizes the development of pest resistance.

A similar middle ground should be reachable between profit-driven corporations and protesters whose positions are not based on the economic bottom line. This compromise position could come about if corporations facilitated technology transfer to third world countries, and if the anticorporate movement accepted the reaping of some profit by corporations doing business in developing countries. As we have seen, corporations recently have reached out through generous royalty-free arrangements that have allowed Golden Rice to proceed toward third world farming markets. In turn, the anti-GM movement might tone down its rhetoric and facilitate the introduction of these important new technologies.

It also is important for aid programs to present diverse options for farmers. GM crops do have a role to play in third world farming, but their overuse would lead to some of the same problems that beset the Green Revolution, which benefited many but not all farmers. Green Revolution varieties and methods were and continue to be useful in many areas, and they have increased food production. But in some regions, farmers cannot afford the high fertilizer and pesticide inputs required by Green Revolution crops, and Green Revolution varieties have not proved viable in all locales.

Similarly, GM crops need to be viewed as plants that will help some but certainly not all producers. We need to maintain and increase access to traditional systems of agriculture, and ensure that historical varieties remain available to those who would benefit more from them than from transgenic crops. Trade agreements being negotiated today contain clauses that provide funds to establish institutions that will promote the preservation and use of traditional varieties. Corporate support and government approval of these provisions would ensure that diverse approaches remain viable options in developing countries.

To ensure our current and future food supplies, both farmers and consumers want sustainable agriculture. "Sustain" is an interesting word, defined as both "to supply with nourishment" and "to affirm the validity of." There is a message here for all the parties to discussions about transgenic crops. Affirming the validity of diverse approaches will be the most successful way of nourishing the human population, and as a first step we need to recognize that organic farmers, conventional growers, corporations, and politically concerned citizens all have important contributions to make to the effort to reach that diversified balance point where all of our best interests are served.

Attitudes toward biotechnology are often driven by strongly held views on the ethics of genetic engineering. For some of us, the ethics of transgenic crops is situational, dependent on context, purpose, and the weighing of benefits and risks. For others, though, it is an absolute issue, with a particular moral stance, political belief, or view about nature leading to outright rejection of anything transgenic.

Some simply believe that genetic engineering goes against divine will. The counter-argument that people have been selecting and breeding livestock and crops for ten thousand years without arousing undue ethical concerns does not resonate with those who cannot countenance biotechnology. For them, genetic engineering is the sole province of the deity. This type of objection is particularly difficult to counter because it arises from the realm of belief, not from the realm of logic.

Nevertheless, science as a discipline has not responded well to moral objections. Scientists usually have chosen to hide behind the subterfuge that genetic engineering differs from previous plant and animal breeding techniques only in the speed and precision with which human-directed selection operates. But the fact is that it is unprecedented to specifically transfer genes from one species to another across the boundaries that until now separated even kingdoms of organisms, and by no stretch of the imagination can doing so be considered equivalent to traditional agricultural practices. The credibility of the scientific community has suffered as a result of its refusal to acknowledge the unprecedented nature of genetic engineering, and the public has been more discerning than scientists in recognizing the magnitude of the choices biotechnology will force us to make. There may well be morally justifiable reasons to proceed with many GM products, but the debate about which of them to pursue would benefit if scientists engaged the ethical issues more directly rather than minimizing the novelty and significance of genetic engineering.

Political true believers also have influenced the GM agenda, with strong opposition coming from those who abhor and fear the influence of large corporations. This has been most prevalent in Britain and continental Europe, where the left-leaning perspective that multinational companies are exploitive, greedy, and out of control is more widespread than in North America. Transgenic agriculture has fallen victim to this deep mistrust, which has engendered concerns about biotechnology that are based not so much on genetic engineering itself as on the corporate source of transgenic crops. For these objectors, commercialization for profit and exploitation of developing countries, rather than panic about the crops themselves, are the primary issues.

Another concern rooted in beliefs more than in data revolves around changes that GM crops might induce in the environment. This concern is particularly strong in Britain, where residents are passionately attached to nature as they know it and fervently opposed to GM crops because genetic drift might alter the contours of the countryside. This segment of the opposition appears to have had enough of human-induced change, which has already radically transformed the landscape.

Opposing this point of view are most farmers, who do not object to modifying nature so long as there are management options available when problems develop. From their perspective, a gene moving from crop to weed is of little concern as long as there is an herbicide that will kill the modified weed in fields, roadsides, parks, or backyards if it becomes a nuisance.

Morality, politics, and perceptions about nature have contributed to the growth of anxieties about GM crops that are unrelated to scientific issues. But such apprehensions that there is just something wrong about genetic engineering play a potent role in the formation of public attitudes toward biotechnology.

We the public have become used to having lobbyists and interest groups asking and answering questions for us. Such advocates may represent legitimate if self-serving and narrow perspectives, but we have become too passive in accepting the constricted opinions they bring to public debate and decision making. Nowhere is this more true than in discussions of science, where the dense jargon and technical terms can make it attractive to let surrogates armed with small and easily digested batches of incomplete information engage issues for us.

Biotechnology is not nearly as difficult to comprehend as experts often maintain, and consumers' interests are more diverse than interest groups admit. Remove the jargon, tone down the rhetoric, and genetically modified crops present comprehensible trade-offs between benefits and risks that are more straightforward than they appear when obscured or misrepresented by the scientists' argot and unidimensional advocacy positions.

The advantages and disadvantages of agricultural genetic engineering can be summarized succinctly. GM crops furnish new products and profits for economic growth, increase crop yields, reduce pesticide use, multiply farming options, and can improve the diet and health of the world's impoverished people. Transgenic crops are not risk free, however. They have the potential to harm natural ecosystems, lead to selection for resistant pests, and cause allergic reactions or other health problems when incorporated in food. They also could be a means through which corporations could exploit all the world's people, especially the poor citizens of developing countries.

None of these problems is intractable, and there is more common ground to be found between differing positions put forth in discussion of GM crops than any participant in the debate has so far admitted. Industry has its humanitarians who are using biotechnology to try to alleviate world hunger; organic farmers share concerns about how mass-production agriculture can feed the growing human population; environmentalists and consumers could support genetic engineering if regulators would furnish reliable reassurances about risks; and even political activists could live with GM crops if international treaties and national policies provided some universal benefits rather than merely facilitating the earning of corporate profits.

Moving toward that middle ground requires two crucial steps. First, the same scientific community that brought us genetic engineering in the first place must provide accurate information about its risks. Government and institutional laboratories could produce exemplary environmental health and food safety evaluations if they were funded and supported with the same level of commitment that characterized the invention and commercialization of GM crops. Accurate, unbiased, comprehensive, and sufficient research studies concerning the risks of transgenic crops and foods is the single most important element missing from the debate so far, and by clarifying risks we would hone our ability to make wise decisions.

Second, our governments need to play a more vital role in creating a broad-based view of genetic engineering that reflects the diverse opinions about this technology rather than simply the narrow perspectives of industry favored by regulators. Governments so far have been more influenced by pressure from proponents to proceed quickly than by public concerns that counsel prudence. Those concerns may well prove to be exaggerated, but if so a government-led increase in regulatory oversight would soon engender greater consumer confidence in transgenic products.

Our current patched-together policies on genetic engineering are satisfying neither industry, which wants to move ahead with new GM crops, nor consumers, who want to know if their food is or is not genetically engineered, nor environmentalists, who fear that transgenic crops may be inflicting irrevocable harm on ecosystems already damaged by human activities. All of these concerns can be dealt with through more broadly based public engagement with the issues, improved government leadership, and an expanded reliance on scientifically independent risk analysis. A public that is well informed about the issues, guided by autonomous science, and led by government regulators with a balanced agenda is likely to become comfortable with most genetically engineered crops.

The potential benefits of biotechnology are too significant, and the risks sufficiently controllable, to make it worthwhile to break the current logjam and move toward consensus on developing and regulating genetically modified crops. As with all compromises, of course, the various sides need to yield points and consider the greater public good more than their own self-interests.

Nowhere in my travels did I encounter a proven risk from transgenic crops sufficiently dangerous to justify stopping their development and sale, but I also did not get the sense during any of my visits that we know enough to proceed without caution. Thorough analysis may indicate that a few next-generation transgenic crops are too risky, and others might require modification to make them field-safe, but most will survive the rigorous scrutiny needed to establish public confidence in GM crops and the food produced from them. Nevertheless, even the most beneficial genetically engineered crops will have some side effects, and we will need to become more willing to accept minor negative impacts if the benefits prove to be substantial.

Science has brought us spectacular innovations coupled with the realization that the undesirable consequences of even the most advantageous discoveries need to be considered and dealt with. This is the crux of the debate about biotechnology—and about all new technologies—finding that balance point between progress and safety, benefit and risk. What is most disturbing about genetic engineering is that this balance point is considerably easier to envision and reach than is apparent from the obfuscatory debates we have experienced so far.

We have made considerably greater progress in achieving scientific breakthroughs than we have in managing the chaotic controversies that swirl around them. The biotechnology experience has revealed a deep chasm between science and public awareness, a divide that will not be bridged until we learn to conduct informed public discussion with the same rigor, creativity, and skill with which we invent new technologies.

Selected References

Prologue

Dobzhansky, Theodosius. 1950. The genetic basis of evolution. *Scientific American* 182:32–41.

James, C. 2000. *Global Review of Commercialized Transgenic crops: 2000.* International Service for the Acquisition of Agri-Biotech Applications, no. 21-2000, Cornell University, Ithaca.

1. Seeds

Brody, H. 2000. *The Other Side of Eden.* Douglas and McIntyre, Toronto.

Darwin, C. 1859. *The Origin of Species.* Oxford University Press, Oxford.

Duvick, D. N. 2001. Biotechnology in the 1930s: the development of hybrid maize. *Nature Reviews Genetics* 2:69–74.

Fowler, C. 1994. *Unnatural Selection.* Gordon and Breach Science Publishers, Yverdon, Switzerland.

Fussell, B. 1992. *The Story of Corn.* North Point Press, New York.

Hancock, J. F. 1992. *Plant Evolution and the Origin of Crop Species.* Prentice Hall, Englewood Cliffs, N.J.

Harlan, J. R. 1992. *Crops and Man.* American Society of Agronomy, Madison, Wisc.

Price, T. D., and A. B. Bebauer. 1995. *Last Hunters—First Farmers.* School of American Research Press, Santa Fe, N.M.

2. In the Heat of the Day

Baltimore, D., and S. Krimsky 1980. The ties that bind or benefit. *Nature* 283:130–131.

Berg, P., D. Baltimore, S. Brenner, R. O. Roblin III, and M. Singer. 1975. Asilomar conference on recombinant DNA molecules. *Science* 188:991–994.

Chalfen, M. H. 1990. Biotechnology facility approval at the local level: The Cambridge, Massachusetts, experience. In *Bioprocessing Safety: Worker and Community Safety and Health Considerations*, ed. W. C. Hyer, Jr. American Society for Testing and Materials, Philadelphia.

Hartley, B. 1980. The bandwagon begins to roll. *Nature* 283:122.

Kornberg, A. 2000. Invention is the mother of necessity. *Toronto Globe and Mail*, 4 November.

Krimsky, S. 1982. *Genetic Alchemy: The Social History of the Recombinant DNA Controversy.* MIT Press, Cambridge, Mass.

Singer, M., and D. Soll. 1973. Guidelines for DNA hybrid molecules. *Science* 181:1114.

Wald, G. 1979. The case against genetic engineering. In *The Recombinant DNA Debate*, ed. David A. Jackson and Stephen P. Stich. Prentice Hall, Englewood Cliffs, N.J.

3. The Regulators

Carrière, Y., T. J. Dennehy, B. Pedersen, S. Haller, C. Ellers-Kirk, L. Antilla, Y. Liu, E. Willott, and B. E. Tabashnik. 2001. Large-scale management of insect resistance to transgenic cotton in Arizona: can transgenic insecticidal crops be sustained? *Journal of Economic Entomology* 94:315–325.

Glare, T. R., and M. O'Callaghan. 2000. *Bacillus thuringiensis: Biology, Ecology, and Safety.* Wiley, New York.

Gould, F. 1988. Evolutionary biology and genetically engineered crops. *BioScience* 38: 26–33.

———2000. Testing Bt refuge strategies in the field. *Nature Biotechnology* 18:266–267.

Gould, F., A. Anderson, A. Jones, D. Sumerford, D. G. Heckel, J. Lopez, S. Micinski, R. Leonard, and M. Laster. 1997. Initial frequency of alleles for resistance to *Bacillus thuringiensis* toxins in field populations of *Heliothis virescens*. *Proceedings of the National Academy of Sciences, U.S.A.* 94:3519–3523.

McGaughey, W. H., F. Gould, and W. Gelernter. 1998. Bt resistance management. *Nature Biotechnology* 16:144–146.

Mellon, M., and J. Rissler. 1998. Now or Never: Serious New Plans to Save a Natural Pest Control. Union of Concerned Scientists, Cambridge, Mass.

Roush, R. T. 1997. Bt-transgenic crops: just another pretty insecticide or a chance for a new start in resistance management? *Pesticide Science* 51:328–334.

Shelton, A. M., J. D. Tang, R. T. Roush, T. D. Metz, and E. D. Earle. 2000. Field tests on managing resistance to Bt-engineered plants. *Nature Biotechnology* 18:339–342.

Tabashnik, B. E. 1994. Evolution of resistance to *Bacillus thuringiensis*. *Annual Review of Entomology* 39:47–79.

Thompson, G. D., and G. Head. 2001. Implications of regulating insect resistance management. *American Entomologist* 47:6–10.

U.S. Environmental Protection Agency. 1999. *EPA and USDA Position Paper on Insect Resistance Management in Bt Crops.* Washington, D.C.

———1999. *EPA/USDA Workshop on Bt Crop Resistance Management, Rosemont, Ill.,* ed. Patrick Steward. Washington, D.C.

———2001. *Bt Plant-Pesticides Risk and Benefit Assessments.* SAP Report no. 2000-07, March 12, Washington, D.C.

U.S. National Research Council. 2000. *Genetically Modified Pest-protected Plants: Science and Regulation.* National Academy Press, Washington, D.C.

4. Of Butterflies and Weeds

Boudry, P., M. Mörchen, P. Saumitou-Laprade, Ph. Vernet, and H. Van Dijk. 1993. The origin and evolution of weed beets: consequences for the breeding and release of herbicide-resistant transgenic sugar beets. *Theoretical and Applied Genetics* 87:471–478.

Brower, L. P. 2001. Canary in the cornfield. *Orion* spring:32–41.

Crawley, M. J., S. L. Brown, R. S. Hails, D. D. Kohn, and M. Rees. 2001. Transgenic crops in natural habitats. *Nature* 409:682–683.

Ellstrand, N. C., H. C. Prentice, and J. F. Hancock. 1999. Gene flow and introgression from domesticated plants into their wild relatives. *Annual Review of Ecology and Systematics* 30:539–563.

Ellstrand, N. C., and K. A. Schierenbeck. 2000. Hybridization as a stimulus for the evolution of invasiveness in plants. *Proceedings of the National Academy of Sciences, U.S.A.* 97:7043–7050.

Fernandez-Cornejo, J., and W. D. McBride. 2000. *Genetically Engineered Crops for Pest Management in U.S. Agriculture: Farm-Level Effects.* Economic Research Service, U.S. Department of Agriculture, Agricultural Economic Report no. 786, Washington, D.C.

Girard, C., A. Picard-Nizou, E. Grallien, B. Zaccomer, L. Jouanin, and M. Pham-Delègue. 1998. Effects of proteinase inhibitor ingestion on survival, learning abilities, and digestive proteinases of the honeybee. *Transgenic Research* 7:1–8.

Hansen-Jesse, L. C., and J. J. Obrycki. 2000. Field deposition of Bt transgenic corn pollen: lethal effects on the monarch butterfly. *Oecologia* 125:241–248.

Hellmich, R. L., and B. D. Siegfried. 2001. *Bt* corn and the monarch butterfly: research update. In *Genetically Modified Organisms in Agriculture,* ed. Gerald C. Nelson. Academic Press, San Diego.

Hilbeck, A., M. Baumgartner, P. M. Fried, and F. Bigler. 1998. Effects of transgenic *Bacillus thuringiensis* corn-fed prey on mortality and development time of immature *Chrysoperla carnea* (Neuoptera: Chrysopidae). *Environmental Entomology* 27:480–487.

Jørgensen, R. B., B. Andersen, A. Snow, and T. P. Hauser. 1999. Ecological risks of growing genetically modified crops. *Plant Biotechnology* 16:69–71.

Losey, J. E., L. S. Rayor, and M. E. Carter. 1999. Transgenic pollen harms monarch larvae. *Nature* 399:214.

Malone, L. A., E. P. J. Burgess, and D. Stefanovic. 1999. Effects of a *Bacillus thuringiensis* toxin, two *Bacillus thuringiensis* biopesticide formulations, and a soybean trypsin inhibitor on honey bee (*Apis mellifera* L.) survival and food consumption. *Apidologie* 30:465–473.

Marvier, M. 2001. Ecology of transgenic crops. *American Scientist* 89:160–167.

Mikkelson, T. B., B. Andersen, and R. B. Jørgenson. 1996. The risk of crop transgene spread. *Nature* 380:31.

Nuffield Council on Bioethics. 1999. *Genetically Modified Crops: The Ethical and Social Issues.* Nuffield Council on Bioethics, London.

Pham-Delègue, M., C. Girard, M. L. Métayer, A. Picard-Nizou, C. Hennequet, O. Pons, and L. Jouanin. 2000. Long-term effects of soybean protease inhibitors on digestive enzymes, survival, and learning abilities of honeybees. *Entomologia Experimentalis et Applicata* 95:21–29.

Picard-Nizou, A. L., M. H. Pham-Delègue, V. Kerguelen, P. Douault, R. Marilleau, L. Olsen, R. Grison, A. Toppan, and C. Masson. 1995. Foraging behaviour of honey bees (*Apis mellifera* L.) on transgenic oilseed rape (*Brassica napus* L. var *oleifera*). *Transgenic Research* 4:270–276.

Pimentel, D., L. Lach, R. Zuniga, and D. Morrison. 2000. Environmental and economic costs of nonindigenous species in the United States. *BioScience* 50:53–65.

Riddick, E. W., and P. Barbosa. 1998. Impact of Cry3A-intoxicated *Leptinotarsa decemlineata* (Coleoptera: Chrysomelidae) and pollen on consumption, development, and fecundity of *Coleomegilla maculaga* (Coleoptera: Coccinellidae). *Annals of the Entomological Society of America* 91:303–307.

Schuler, T. H., R. P. J. Potting, I. Denholm, and G. M. Poppy. 1999. Parasitoid behaviour and Bt plants. *Nature* 400:825.

Snow, A. A., B. Andersen, and R. Bagger Jørgensen. 1999. Costs of transgenic herbicide resistance introgressed from *Brassica napus* into weedy *B. rapa*. *Molecular Ecology* 8:605–615.

Snow, A. A., and P. M. Palma. 1997. Commercialization of transgenic plants: potential ecological risks. *BioScience* 47:86–97.

Timmons, A. M., E. T. O'Brien, Y. M. Charters, S. J. Dubbels, and M. J. Wilkinson. 1995. Assessing the risks of wind pollination from fields of genetically modified *Brassica napus* ssp. *oleifera. Euphytica* 85:417–423.

Wolfenbarger, L. L., and P. R. Phifer. 2000. The ecological risks and benefits of genetically engineered plants. *Science* 290:2088–2093.

Wraight, C. L., A. R. Zangerl, M. J. Carroll, and M. R. Berenbaum. 2000. Absence of toxicity of *Bacillus thuringiensis* pollen to black swallowtails under field conditions. *Proceedings of the National Academy of Sciences, USA,* 97:7700–7703.

5. It Only Moves Forward

Aventis. 2000. Updated safety assessment in support of the pesticide petition for a time-limited exemption from the requirement of a tolerance for the plant pesticide *Bacillus thuringiensis* subsp. *tolworthi* Cry9C. Submission to the U.S. Environmental Protection Agency, registration number 264–669.

Consumers' Association of Canada. 1999. Statement on Biotechnology. Ottawa, 2 November.

Gaskell, G., M. W. Bauer, J. Durant, and N. C. Allum. 1999. Worlds apart? the reception of genetically modified foods in Europe and the U.S. *Science* 285:384–387.

GeneWatch. 1998. *Genetically Modified Foods: Will Labeling Provide Choice?* Briefing no. 1. Buxton, Derbyshire, Eng.

Goldman, K. A. 2000. Bioengineered food-safety and labeling. *Science* 290:458–459.

McGarity, T. O., and P. I. Hansen. 2001. *Breeding Distrust: An Assessment and Recommendations for Improving the Regulation of Plant Derived Genetically Modified Foods.* Food Policy Institute, Consumer Federation of America, Washington, D.C.

Royal Society of Canada. 2001. *Elements of Precaution: Recommendations for the Regulation of Food Biotechnology in Canada.* Ottawa.

U.S. Centers for Disease Control and Prevention. 2001. *Investigation of Human Health Effects Associated with Potential Exposure to Genetically Modified Corn.* Atlanta, 11 June.

U.S. Environmental Protection Agency. 2000. *Assessment of Scientific Information concerning StarLink™ Corn.* SAP Report no. 2000-06, 1 December, Washington, D.C.

————2001. *FDA Evaluation of Consumer Complaints Linked to Foods Allegedly Containing StarLink™ Corn.* Washington, D.C., 13 June.

————2001. Voluntary labeling indicating whether foods have or have not been developed using bioengineering. Center for Food Safety and Applied Nutrition, FDA, docket Nnmber 00D-1598, Washington, D.C.

6. Saving the Family Farm

Canola Council of Canada. 1999. *Canola Production Centre 1999 Report.* Winnipeg, Manitoba.

Downey, R. K., Jr. 2000. Canola—A Canadian Success Story. Unpublished manuscript.

National Research Council of Canada. 1992. *From Rapeseed to Canola: The Billion Dollar Success Story.* Plant Biotechnology Institute, Saskatoon, Saskatchewan.

Palmer, C. E., and W. A. Keller. 2000. Transgenic Oilseed Brassicas. Unpublished manuscript.

Smith, M. 1999. Food fright. *Toronto Globe and Mail* report on *Business Magazine,* November.

7. Saving the Bugs

Browning, F. 1998. *Apples.* Farrar, Straus & Giroux, New York.

Canadian General Standards Board. 1999. *Organic Agriculture.* CAN/CGSB-32.310.99. Ottawa.

Clarke, M. 2000. *Draft Report and Recommendations, GMO Task Force.* Organic Trade Association, Greenfield, Mass.

Edwards, L. 1998. *Organic Tree Fruit Management.* Certified Organic Associations of British Columbia, Keremeos, British Columbia.

Organic Trade Association. 1999. *American Organic Standards.* Greenfield, Mass.

U.S. Department of Agriculture. 2000. Proposed Rule, National Organic Program. Docket Number TMD-00-02-PR2, Agricultural Marketing Service, USDA, Washington, D.C.

8. Anything under the Sun

Barton, J. H. 1998. The impact of contemporary patent law on plant biotechnology research. In *Intellectual Property Rights III Global Genetic Resources: Access and Property Rights.* Crop Science Society of America, Madison, Wisc.

———2000. Reforming the patent system. *Science* 287:1933–1934.

Cleveland, D. A., and S. C. Murray. 1997. The world's crop genetic resources and the rights of indigenous farmers. *Current Anthropology* 38:477–515.

Dawkins, K. 1999. Intellectual property rights and the privatization of life. *Focus Magazine* 4:1–7.

Enriquez, J., and R. A. Goldberg. 2000. Transforming life, transforming business: the life science revolution. *Harvard Business Review* March–April:96–104.

Erbisch, F. H., and C. Velazquez. 1998. *Intellectual Property Rights in Agricultural Biotechnology.* CABI International, Wallingford, Eng.

Esquinas-Alcázar, J. 1998. Farmers' rights. In *Agricultural Values of Plant Genetic Resources,* ed. R. E. Evenson, D. Gollin, and V. Santaniello. CABI Publishing, Wallingford, Eng.

Fowler, C. 1994. *Unnatural Selection.* Gordon and Breach Science Publishers, Yverson, Switzerland.

Heller, M. A., and R. S. Eisenberg. 1998. Can patents deter innovation? the anticommons in biomedical research. *Science* 280:698–701.

Jasanoff, S. 1995. *Science at the Bar.* Harvard University Press, Cambridge, Mass.

Thompson, P. B. 1995. Conceptions of property and the biotechnology debate. *BioScience* 45:275–282.

———1997. *Food Biotechnology in Ethical Perspective.* Blackie Academic and Professional, London.

9. There'll Always Be an England

Dibb, S., and S. Mayer. 2000. *Biotech—The Next Generation: Good for Whose Health.* The Food Commission, London.

GeneWatch. 1998. Genetically engineered crops and food: The case for a moratorium. Briefing no. 4. GeneWatch, Buxton, Derbyshire, Eng.

———1998. *Genetically Engineered Oilseed Rape: Agricultural Saviour or New Form of Pollution?* Briefing no. 2. GeneWatch, Buxton, Derbyshire, Eng.

———1999. *Farm Scale Trials of GM Crops: Answering the Safety Questions?* Briefing no. 8. Buxton, Derbyshire, Eng.

———1999. *Genetic Engineering: A Review of Developments in 1998.* Briefing no. 5. Buxton, Derbyshire, Eng.

———2000. *GM Crops and Food: A Review of Developments in 1999.* Briefing no. 9. Buxton, Derbyshire, Eng.

Homer-Dixon, T. 2000. *The Ingenuity Gap.* Alfred A. Knopf Canada, Toronto.

Nuffield Council on Bioethics. 1999. *Genetically Modified Crops: The Ethical and Social Issues.* Nuffield Council on Bioethics, London.

Paxman, J. 1998. *The English: A Portrait of a People.* Penguin Books, London.

10. For the Good of Mankind

Dibb, S., and S. Mayer. 2000. *Biotech—The Next Generation: Good for Whose Health.* The Food Commission, London.

Kryder, R. D., S. P. Kowalski, and A. F. Krattiger. 2000. International Service for the Acquisition of Agri-biotech Applications, Brief no. 20, Ithaca.

Paarlberg, R. L. 2000. Genetically modified crops in developing countries: promise or peril? *Environment* 42:19–27.

———2000. The global food fight. *Foreign Affairs* 79:24–38.

———2001. *The Politics of Precaution: Genetically Modified Crops in Developing Countries.* Published by Johns Hopkins University Press for the International Food Policy Research Institute, Baltimore.

Runge, C. F., and B. Senauer. 2000. A removable feast. *Foreign Affairs* 79:39–51.

Shewmaker, C. K., J. A. Sheehy, M. Daley, S. Colburn, and D. Y. Ke. 1999. Seed-specific overexpression of phytoene synthase: increase in carotenoids and other metabolic effects. *The Plant Journal* 20:401–412.

Shiva, V. 2000. *Vit A Rice a Blind Approach to Blindness Control.* Research Foundation for Science, Technology, and Ecology, 14 February, New Delhi, 14 February.

Ye, X., S. Al-Babili, A. Klöti, J. Zhang, P. Lucca, P. Beyer, and I. Potrykus. 2000. Engineering the provitamin A (ß-carotene) biosynthetic pathway into (carotenoid-free) rice endosperm. *Science* 287:303–305.

11. Risks Real or Imagined

Carlson, R. 2000. The world in 2050. The Economist/Shell World in 2050 Essay Competition.

Acknowledgments

It is once again a deep pleasure to acknowledge my editors at Harvard University Press for sharing their passion for books and their expertise in the craft of writing. I particularly value my friendship and collaboration with Michael Fisher, a consummate critic, engaged observer, and master of words who has immeasurably improved this book and made *Travels* a delightful project. I also am grateful to Nancy Clemente, whose fine editorial touch substantially improved the manuscript, and to Annamarie Why for her book design.

I also thank the many people who sent manuscripts, books, and articles for me to peruse, provided innumerable contacts, and agreed to take the time to talk with me about biotechnology. I did not in the end agree with all their opinions, but I did value their candor, respect their perspectives, appreciate their passion and vision, and above all enjoy each of their sometimes-quirky but always interesting personalities. In particular, I would like to thank:

Chapter 1: Von Kaster kindly arranged a day at Garst Seeds, and I appreciated his insights as well as those of Alan Hawkins and Ben Hable about the company's past, present, and future. Candy Gardner from the U.S. Department of Agriculture Plant Introduction Station at Iowa State University in Ames arranged a tour of the facilities and shared her views on the preservation of agricultural biodiversity. Donald Duvick, now retired but formerly employed by Pioneer Hi-Bred, provided background information on hybrid corn. I also am grateful to Liz Garst, David Garst, and the staff at the Garst Farm Resort for a pleasant and relaxing weekend and for their reminiscences about Roswell Garst.

Chapter 2: I thank Dean Bushey and Alison Chalmers of Aventis and Rich Lotstein of Syngenta for talking to me about their work in industry and sharing their opinions about industrial science. Ed Vargo from North Carolina State University efficiently arranged my visit to Research Triangle Park. I also thank Melvin Chalfen, now retired but active in the Cambridge, Massachusetts, Public Health Department during the recombinant DNA days, who provided useful background information about the period and jogged many memories.

Chapter 3: I am deeply grateful to Fred Gould, who guided me through the intricacies of the personalities and procedures of biotechnology regulation, and as always was a delight to talk with. I also appreciated the time I spent conversing with Jane Rissler, Rebecca Goldburg, and Doreen Stabinsky from the Union of Concerned Scientists, the Environmental Defense Fund, and Greenpeace respectively. Some background information was provided by Janet Andersen, Janet Carpenter, Brian Federici, Dick Hardee, and Bruce Tabashnik.

Chapter 4: Rick Hellmich from the USDA Department of Agriculture in Ames, Iowa, shared much useful information and opinions, as did Karen Oberhauser of the University of Minnesota and Orley Taylor of the University of Kansas. I am also grateful to John Losey of Cornell University and May Berenbaum of the University of Illinois for talking candidly with me about their experiences in the butterfly wars.

Chapter 5: The Consumer Association of Canada's Jenny Hillard allowed me to sit in on one of her focus group sessions, and also shared with me her passion for consumer rights.

Chapter 6: I am deeply grateful to Don and Jamie Dixon for their friendship and warm hospitality during a cold prairie winter. JoAnne Buth and Dale Adolphe from the Canola Council of Canada were unusually helpful in suggesting people to interview and giving me their views. I also appreciated my conversations with farmers: Wayne Bacon, Roy Button, Bruce Dalgarno, Darren Qualman, Jonothon Roskos, Percy Schmeiser, and Ray Wilfing. Wilf Keller from Agriculture Canada's Plant Biotechnology Institute (PBI) spent an afternoon sharing his enthusiasm for the science and its prospects, and Zamir Punja of Simon Fraser University, Brian Ellis of the University of British Columbia, and Keith Downey of PBI answered many questions for me.

Chapter 7: I appreciate the assistance of Linda Edwards, an organic grower from British Columbia, who suggested several excellent directions for me to pursue. I talked with, received information from, and enjoyed the company of many organic farmers and food brokers, especially Margaret Clark, Mariah Cornwoman, Gene Kahn, Linda Lutz, Harold Ostenson, Phil Unterschuetz, Chuck Walker, and Roger Wechsler. I also was grateful to be invited to attend an excellent organic banquet put on by Debby Boyle of ProOrganics in Burnaby, British Columbia.

Chapter 8: John Doll of the U.S. Patent and Trademark Office was one of the most refreshing, direct, and candid people I met during my travels, and what I learned from him was instrumental in shaping the discussion of patents and intellectual property. I also spent a wonderful day with Margaret Llewelyn of Sheffield University learning about the British point of view on patent issues.

Chapter 9: I am deeply grateful to Sue Mayer of GeneWatch for helping to arrange my trip to Britain, sharing her balanced and discerning thoughts about biotechnology, and taking me on a delightful walk through the English countryside. Vernon Barber, Jayn Harding, Stuart Thomson, and Rachel Wilson also provided me with useful material, and Charles Mann kindly read the manuscript of this chapter and suggested useful changes at the beginning and end.

Chapter 10: I appreciate the assistance of Eric Sachs of Monsanto, who opened the corporate doors and was a candid and pleasant host. The book benefited immeasurably from his efforts and from discussions with Cherian George, Robert Horsch, Bryan Hurley, Maureen Mackay, and George Michaels. Dave Kryder of ISAAA helped me understand the corporate perspective on developing countries. I also am grateful to Rob Paarlberg, who not only spent many hours discussing biotechnology and the third world with me, but also sent me a copy of the manuscript of his forthcoming book.

My colleague John Borden and all of our students engaged me in always-stimulating lunchtime discussions, and encouraged me to press on when my energy flagged. I also appreciate the cyberstream of articles and relevant Web sites sent by Glenn Bullard and Ruth and Larry Winston, and the able library research conducted by Sarah Butler, which saved me many hours of trekking across campus.

I would like to acknowledge financial support from the Canada Council of the Arts. Although the council did not directly support this project, the book did benefit indirectly from the Killam Fellowship awarded me by the council and from the invitation extended to me in August 2000 to read at the Sunshine Coast Writers Festival, funded in part by the council.

Finally, I am deeply grateful to my wife, Susan Katz, who listened to endless monologues, read every sentence of the manuscript, and provided unflagging encouragement.

Index